IEA Research for Education

A Series of In-depth Analyses Based on Data
of the International Association for the Evaluation
of Educational Achievement (IEA)

Volume 14

The International Association for the Evaluation of Educational Achievement (IEA) is an independent nongovernmental nonprofit cooperative of national research institutions and governmental research agencies that originated in Hamburg, Germany in 1958. For over 60 years, IEA has developed and conducted high-quality, large-scale comparative studies in education to support countries' efforts to engage in national strategies for educational monitoring and improvement.

IEA continues to promote capacity building and knowledge sharing to foster innovation and quality in education, proudly uniting more than 60 member institutions, with studies conducted in more than 100 countries worldwide.

IEA's comprehensive data provide an unparalleled longitudinal resource for researchers, and this series of in-depth peer-reviewed thematic reports can be used to shed light on critical questions concerning educational policies and educational research. The goal is to encourage international dialogue focusing on policy matters and technical evaluation procedures. The resulting debate integrates powerful conceptual frameworks, comprehensive datasets and rigorous analysis, thus enhancing understanding of diverse education systems worldwide.

Nani Teig · Trude Nilsen · Kajsa Yang Hansen
Editors

Effective and Equitable Teacher Practice in Mathematics and Science Education

A Nordic Perspective Across Time
and Groups of Students

Editors
Nani Teig
Department of Teacher Education
and School Research
University of Oslo
Oslo, Norway

Trude Nilsen
Department of Teacher Education
and School Research, CREATE—Centre
for Research on Equality in Education
University of Oslo
Oslo, Norway

Kajsa Yang Hansen
Department of Education and Special
Education
University of Gothenburg
Gothenburg, Sweden

ISSN 2366-1631   ISSN 2366-164X (electronic)
IEA Research for Education
ISBN 978-3-031-49579-3   ISBN 978-3-031-49580-9 (eBook)
https://doi.org/10.1007/978-3-031-49580-9

In loving memory of Prof. Sigrid Blömeke

Your passion was a glowing flame, brightening every discussion and igniting inspiration within us. You taught us to question, to explore, and to never settle for the easy answers. This book is a tribute to you, an effort to advance the conversation on educational effectiveness and equity that you so passionately advocated for. Although your journey has come to an end, the paths you carved continue to guide and inspire us, now and always.

We remember, we honour, and we continue the quest.

Series Editors' Foreword

IEA's mission is to enhance knowledge about education systems worldwide and to provide high-quality data that will support education reform and lead to better teaching and learning in schools. In pursuit of this aim, it conducts, and reports on, major studies of student achievement in literacy, mathematics, science, citizenship, and digital literacy. These studies, most notably TIMSS, PIRLS, and ICCS, are well established and have set the benchmark for international comparative studies in education.

The studies have generated vast datasets encompassing student achievement, disaggregated in a variety of ways, along with a wealth of contextual information which contains considerable explanatory power. The numerous reports that have emerged from them are a valuable contribution to the corpus of educational research.

Valuable though these detailed reports are, IEA's goal of supporting education reform needs something more: deep understanding of education systems and the many factors that bear on student learning advances through in-depth analysis of the global datasets. IEA has long championed such analysis and facilitates scholars and policy makers in conducting secondary analysis of our datasets. So, we provide software such as the International Database Analyzer to encourage the analysis of our datasets, support numerous publications including a peer-reviewed journal—*Large-Scale Assessments in Education*—dedicated to the science of large-scale assessment and publishing articles that draw on large-scale assessment databases. We also organize a biennial international research conference to nurture exchanges between researchers working with IEA data (https://www.iea.nl/our-conference).

The *IEA Research for Education* series represents a further effort by IEA to capitalize on our unique datasets, so as to provide powerful information for policy makers and researchers. Each report focuses on a specific topic and is produced by a dedicated team of leading scholars on the theme in question. Teams are selected on the basis of an open call for tenders; there are two such calls a year. Tenders are subject to a thorough review process, as are the reports produced. (Full details are available on the IEA website).

The 14th volume in this series focuses on teacher practice, specifically in mathematics and science at grade 4, across the Nordic countries. In producing it, we are delighted to be joined by the Nordic Council of Ministers, as we were for Volume 11 on civics and citizenship education (https://link.springer.com/book/10.1007/978-3-030-66788-7).

It is easy to assert that good teaching is key to powerful learning by students. What this means in practice is not so straightforward, however. What does good teaching—especially for diverse groups—mean? How do we break it down into its component parts? And how do the different elements of good teaching relate to equity considerations? How do we measure student learning across different systems and schools? And how do we relate teacher practice to student learning, and in a way that takes account, for instance, of students' socioeconomic status?

The authors have drawn on Trends in Mathematics and Science Study (TIMSS) data from four Nordic countries, over three cycles and eight years, to zone in on specific aspects of teacher practice and how they relate to the familiar TIMSS measures of student achievement in mathematics and science at grade 4. They also draw on student socioeconomic status and ethnic minority membership data to build up a picture of equity (or lack of) in accessing educational opportunities. In line with the Nordic model of comprehensive schooling and aspiration toward equitable provision, this enables an account, over time, of teacher practice in relation to both effectiveness and equity.

Some findings in the report will be challenging to Nordic policy makers, teacher educators, and teachers. If uncomfortable facts are not faced, however, both student achievement and social equity will continue to be compromised. By delving into the elements of teacher practice and relating them to student achievement and characteristics, the report points to ways forward for school practices and teacher education that have resonance far beyond the Nordic community. It also makes an important contribution to current debates on the tensions between the dominant standards agenda and the ideal of an inclusive school where every student has access to appropriate, high-quality education.

Future volumes in the series include one dedicated to examining the relationship between socioeconomic segregation between classrooms and student outcomes and another volume focusing on the Dinaric region that presents a collection of analyses of reading literacy factors.

Coventry, UK   Seamus Hegarty
Bloomington, USA   Leslie Rutkowski
                                                                                                             Series Editors

Foreword

Nordic Council of Ministers (NCM) is the official body for collaboration between the Nordic governments. Over the years, this intergovernmental co-operation has been a platform for generating synergies, removing obstacles for regional collaboration and mobility, and for sharing knowledge and experiences of a wide range of issues, to benefit the citizens of the region.

Currently, NCM is working toward its Vision 2030 to make the Nordic region the most sustainable and integrated region in the world by 2030. The strategic priorities for co-operation are based on the vision of a green, competitive, and socially sustainable Nordic region.

Quality education and research are central prerequisites to reach this vision. In times of growing inequalities in education, we need to further our knowledge about teaching and learning in order to maintain the Nordic model of education and enforce its ability to work for equity, equal opportunity, and inclusion.

Qualitative international comparable data is essential for evidence-based policy making. For this reason, NCM has been running a series of publications named Northern Lights that compares education within Nordic countries based on large-scale international studies on school performance and teaching. We are delighted to work once again with the International Association for the Evaluation of Educational Achievement (IEA), after establishing a fruitful co-operation in the first edition of this series. This work is also supported by a Nordic expert group consisting of representatives from the national organizations of educational evaluation.

This current edition focuses on a Nordic secondary analysis on the Trends in International Mathematics and Science Study (TIMSS). The aim is to gain new insights on the development of teaching and learning over time within the school subjects of mathematics and science, and with a special focus on teaching practices.

Our sincere hope is that this book provides food for thought in the continued dialogue among a variety of stakeholders on how to best equip our children and youth with relevant competencies in mathematics and science in a world calling for solutions for sustainability.

I also want to express my warmest thanks for the excellent collaboration with the Nordic expert group coordinated by chief adviser Hjalte Meilvang at the Danish Agency for Education and Quality, IEA, series editors Seamus Hegarthy and Leslie Rutkowski as well as to the editors of this book, associate professor Nani Teig and research professor Trude Nilsen from the University of Oslo, and professor Kajsa Yang Hansen from the University of Gothenburg.

Copenhagen, Denmark   Karen Ellemann
                                                                Secretary General
                                                      Nordic Council of Ministers

Contents

Chapter 1
Introduction: Student Achievement and Equity Over Time in the Nordic Countries

Christian Christrup Kjeldsen⑩, **Trude Nilsen**⑩, **Jenna Hiltunen**⑩, and **Nani Teig**⑩

1.1 Introduction

This book examines teacher practices in primary school by exploring the content teachers cover in their teaching, the quality of their teaching, and their assessment practices. The aim is to examine how these practices are related to achievement and changes in achievement in mathematics and science over time, as well as how they are related to educational equity in the Nordic countries. Hence, it is important to provide a backdrop for the book with regard to student achievement, the changes in these achievements, and the inequalities in outcomes over time, in the Nordic countries.

C. C. Kjeldsen (✉)
The Danish School of Education, Aarhus University, Jens Chr. Skous Vej 4, Nobelparken, Building 1483, Room 619, 8000 Aarhus C, Denmark
e-mail: kjeldsen@edu.au.dk

T. Nilsen
Department of Teacher Education and School Research, CREATE—Centre for Research on Equality in Education (project number 331640), University of Oslo, Postboks 1099, Blindern, 0317 Oslo, Norway
e-mail: trude.nilsen@ils.uio.no

J. Hiltunen
University of Jyväskylä, Finnish Institute for Educational Research, P.O. Box 35, 40014 Jyväskylä, Finland
e-mail: jenna.hiltunen@jyu.fi

N. Teig
Department of Teacher Education and School Research, University of Oslo, Postboks 1099, Blindern, 0317 Oslo, Norway
e-mail: nani.teig@ils.uio.no

© The Author(s) 2024
N. Teig et al. (eds.), *Effective and Equitable Teacher Practice in Mathematics and Science Education*, IEA Research for Education 14,
https://doi.org/10.1007/978-3-031-49580-9_1

1

The purpose of this chapter is twofold: (1) to provide the rationale and aims of the book along with an overview of the subsequent chapters, and (2) to establish the necessary backdrop for the book. Section 1.2 outlines the rationale and aim of the book along with a summary of the chapters. Section 1.3 describes the backdrop in terms of achievements and inequalities, comprising a theoretical foundation for equity (Sect. 1.3.1), and three empirical sections (Sects. 1.3.2, 1.3.3, and 1.3.4). Section 1.3.2 presents student achievements and standard deviations, Sect. 1.3.3 examines inequalities among students from minority groups, and Sect. 1.3.4 provides results on inequalities related to students' socioeconomic backgrounds. All empirical sections provide results for the Nordic countries over time. The concluding remarks are presented in the final section (Sect. 1.4).

1.2   The Rationale of the Book and Overview of Chapters

The rationale behind the book lies in the need for more research on teacher practices in the Nordic context. Despite the importance of teacher practice in facilitating effective and equitable learning (Kyriakides et al., 2018; Wahlström, 2022), most research examining different factors of teacher practice and their relationships to student outcomes has taken place outside of the Nordic countries, often in Germany or the United States, and are predominantly focused on secondary education (e.g., Fauth et al., 2021; Kane & Staiger, 2012; Praetorius et al., 2017; Schmidt et al., 2008). Consequently, there is a gap in research in this field for primary schools in the Nordic countries. Accordingly, in order to maintain the Nordic model of education where equity, equal opportunity, and inclusion play critical roles in schooling (Frønes et al., 2020), the Nordic countries need further investigation into these key educational factors. Better knowledge in these areas may strengthen policy and practice to better face the school and societal challenges of today and the future. Such knowledge is even more urgent than ever before, due to the assumed recent increase in educational inequalities caused by the COVID-19 pandemic. To this end, school subjects within the science, technology, engineering, and mathematics (STEM) field, are especially important in preparing students for challenges and opportunities for sustainable development.

To fill this gap in the Nordic educational policy and practice, but also within research in general, it is crucial to understand what makes a teacher practice effective and equitable in mathematics and science education. In this regard, in this book, we focus on *what teachers teach* (content coverage), *how teachers teach* (teaching quality), and *how teachers assess their students* (assessment practice). Specifically, it is crucial to understand to what extent teaching quality and assessment practices, as well teaching content, have changed over time, which aspects are related to student learning outcomes and educational equity, and whether changes in these practices are related to changes in student learning outcomes over time. Such knowledge may also contribute to addressing the inequalities in the development of student mathematics and science learning outcomes in the Nordic countries. Increased knowledge in these areas could potentially bring the countries closer to a more unified Nordic model of education in terms of equitable outcomes.

The overall aim of the book is hence to examine how teacher practices change over time, how they are related to student outcomes, whether changes in teacher practices are related to changes in achievement, and how teacher practices are related to equity in the Nordic countries. To address this aim and fill the existing knowledge gap related to the Nordic countries and primary school education, this book uses data from IEA's (International Association for the Evaluation of Educational Achievement) Trends in Mathematics and Science Study (TIMSS), in which the Nordic countries participated. TIMSS measures students' competence in mathematics and science according to the participating countries' curricula in grades four and eight. TIMSS is a trend study conducted every four years, enabling comparisons over time. While all Nordic countries (except for Iceland) participated in the study for grade four, only two countries participated in the study for grade eight between 2011 and 2019 (Mullis et al., 2020). Hence, in order to include as many Nordic countries as possible and to consider the lack of research in primary school settings, we selected grade four for analysis. Furthermore, TIMSS has representative samples of students and is the only international large-scale assessment (ILSA) that includes intact classrooms of students and teacher questionnaires, linking each student and classroom to their respective teacher (Martin et al., 2020). Given these reasons, TIMSS is deemed the most suitable ILSA for studying teacher practices in the Nordic context (for more information on TIMSS, see Chap. 3).

To study teacher practices over time, the book focuses on the cycles of TIMSS conducted in 2011, 2015, and 2019. These last three cycles provide the most comparable measures of contextual information, such as teacher practices and student characteristics. Earlier cycles were not included due to potential changes in the measures that could compromise the validity of inferences.

The structure of this book is illustrated in Fig. 1.1 and follows the main aims described above. While the present chapter establishes the contexts of the Nordic countries in terms of achievements and equity, Chap. 2 provides the theoretical foundation of the book by describing the conceptualizations of teacher practices and reviewing previous research. Chapter 3 describes the methodological aspects of the book, including the design and frameworks of TIMSS as well as the analytical methods employed.

As illustrated in Fig. 1.1, the book organizes the empirical chapters according to three approaches. The first approach, "relations to outcome in 2019," investigates the relations between teacher practices and student achievements in mathematics and science, as well as the changes in the *means* of the variables measuring teacher practices over time. This section focuses on content coverage (Chap. 4) and teaching quality and assessment practices (Chap. 5). The second approach, "explaining changes in achievement over time," investigates whether changes in teacher practices are related to changes in achievement from 2011 to 2019. This approach is used to investigate content coverage (Chap. 6) and teaching quality

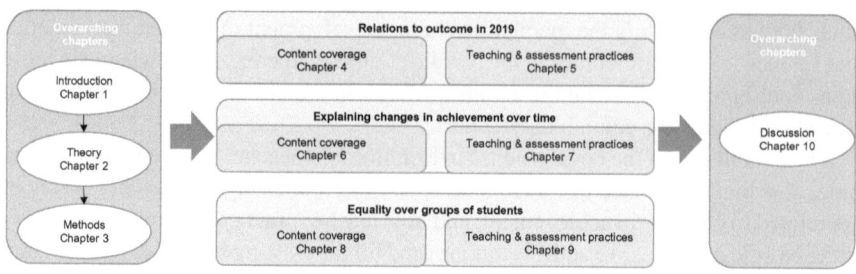

Fig. 1.1 Structure of the book according to three empirical approaches

and assessment practices (Chap. 7). In the third approach, "equality over groups of students", the authors examine how teacher practices are related to inequity. This analysis is used to investigate content coverage (Chap. 8) and teaching quality and assessment practices (Chap. 9). The final chapter serves as an overarching chapter that discusses the findings from the empirical chapters (Chap. 10).

Seeing how the overarching assumption of the book rests on the interconnection between teacher practices and achievement and equity, it is important to provide information about the current state as well as changes in achievement and equity over time in the Nordic countries. This backdrop facilitates the interpretations of findings from the chapters and is presented in Sect. 1.3.

1.3   Achievements and Inequity in Achievements Over Time

This section starts by discussing theoretical perspectives on equity in education, including a discussion on the Nordic model of education (Sect. 1.3.1). It is followed by three empirical subsections that present results on achievements and standard deviations over time (Sect. 1.3.2), the relation between student minority status and achievement over time (Sect. 1.3.3), and the link between socioeconomic status (SES) and achievement over time (Sect. 1.3.4).

1.3.1   Theoretical Considerations on Equity in Education

Generally, the term *educational equity* denotes the provision of equal opportunities for all students, irrespective of their SES, gender, ethnicity, cultural background, or cultural capital, as defined by Bourdieu and Wacquant (2002). It embodies the principle that pedagogical practices and policies should strive to eliminate disparities in educational access, motivation, sense of belonging, well-being, and, importantly, learning outcomes among students from diverse backgrounds. The principle of educational equity therefore advocates for equal opportunities for all students to reach their full potential, regardless of their sociodemographic circumstances.

The understanding of social inequity is intricately linked to various normative considerations concerning the demarcation between just and unjust disparities in society. Based on a synthesis of the different discussions on the concept of equity (see e.g., Espinoza, 2007; Hansen, 1973; Pedersen, 2014; Sen, 1980, 2008; UNESCO, 2015), it can be argued that achieving fair and socially just educational equity at a societal level necessitates a commitment to consistent equity when addressing learning outcomes across groups of students characterized by various external factors, such as SES, gender, language spoken at home, or ethnic origin. By comprehending and addressing the root causes of inequity, policymakers and educators can endeavour to create a more equitable and just educational system.

Educational equity is inherently linked to matters of social justice and human rights, grounded in the belief that education is a public good that benefits both individuals and society at large. From the vantage point of individual learners, access to high-quality education not only pertains to their personal growth but also contributes to fostering a more inclusive and equitable society. Moreover, this approach may potentially enhance a society's economic growth and social cohesion (Burroughs et al., 2019; OECD, 2012).

According to the OECD (2012) and their findings from the Programme for International Student Assessment (PISA), effective educational institutions require a proper blend of skilled staff, sufficient educational assets, and motivated learners, with resources allocated accordingly. The OECD's findings indicate that disadvantaged students often attend schools with limited resources, impacting various aspects of education. Thus, equitable distribution of resources must be considered in school systems (OECD, 2012). As Esping-Andersen (2008) formulates, "The point is that welfare and efficiency concerns coincide. From an equity perspective, children's life chances should depend less on the lottery of birth than on their own latent abilities" (p. 23). Skilled teachers and school resources are pivotal to the learning opportunities of the individual. As such, teachers play a substantive key role in whether *all* students experience equal opportunities to learn. Still, the question of equal opportunities for all is a composite concept. Do equal opportunities also imply equal outcomes for all students despite diversity in their characteristics, interests, effort, or innate potentials? Or is it a question of providing the same opportunities for all? In this matter, we would argue for the importance of the concept of educational equity in decision-making and teacher practice, while additionally, recognizing the challenges involved in achieving it. Nevertheless, schools face different obstacles in fulfilling this ambition.

A focus on equity implies that a specific allocation must be substantiated through a combination of references to abstract principles and tangible evidence. Equity entails examining the social justice implications of education, particularly the fairness, justness, and impartiality in the distribution of resources and opportunities across all levels or sectors of education. In this context, equity is understood as the fair or justified allocation of resources or opportunities (UNESCO Institute for Statistics, 2018).

The Nordic Educational Model

The focal Nordic countries in this book, comprising Denmark, Finland, Norway, and Sweden, are widely recognized for their extensive welfare systems that emphasize equal opportunities and social justice, contributing significantly to their development. These countries frequently receive recognition for their dedication to social justice and equity, particularly within the realm of education. This commitment is most evident during the classical period of the Nordic educational model, which spans from the post-World War II era until the 1970s. During this period, "the main objective was to involve the school in the realisation of social goals such as equal opportunity and community fellowship" (Telhaug et al., 2006, p. 245). Consequently, one might anticipate strong educational equity within the Nordic countries. Nevertheless, a comparative analysis of these countries reveals opportunities for further improvement in achieving this overarching ambition (see Frønes et al., 2020; Nilsen et al., 2018; OECD, 2019).

Within the Nordic welfare state, diverse forms of equity have been explicitly pursued as political objectives. This normative ambition materializes in numerous manifestations, but as Hansen (1986) contends:

> The Scandinavian welfare state has not only searched for the elimination of poverty, but the decrease of inequality as well. When economic growth policy and the struggle against mass poverty are combined with the demand for equality, one finds the combined elements in a social development policy that justify talk of a Scandinavian development model in the form of a vision as an ideological driving force. (p. 109)

The overall emphasis on equity in the Nordic countries described by Hansen is well in line with the developments of a Nordic educational model. The Nordic education model, which sought to balance equity and excellence for all children, arguably peaked between the 1960s and 1980s. During this post-war era and up to the 1980s, Nordic countries were guided by social democratic parties advocating for a welfare state that prioritized equity, justice, and democracy (Tröhler et al., 2022). In the time period examined in this book (2011–2019), the Nordic countries continued to prioritize educational equity. However, the spirit, focus, and emphasis on equity have decreased since the model's peak, due to a number of factors (Frønes et al., 2020). For instance, Sweden implemented free school choice in the early 1990s (Yang Hansen & Gustafsson, 2019), and there have been shifts in the policy and with accumulating evidence of decreased equity in several of the Nordic countries (Yang Hansen & Gustafsson, 2019; Nilsen et al., 2018; OECD, 2019). It is thus important to provide an overview of the development of equity in the Nordic countries over time, as explored in the three following empirical subsections.

1.3.2 Results: Student Achievement and Standard Deviation in the Nordic Countries

Historically, the claim of an equity-efficiency trade-off, in which improvements in overall student achievement come at the expense of a more equitable distribution of educational resources, has been a topic of debate. However, contemporary empirical evidence challenges this idea. In fact, recent findings demonstrate a positive correlation between enhanced educational equity and higher average student performance (e.g., Burroughs et al., 2019; Kyriakides et al., 2018).

Figures 1.2 and 1.3 illustrate the changes in mean achievement for mathematics and science across the TIMSS cycles (2011, 2015, and 2019) in the Nordic countries (excluding Iceland). These figures depict both significant positive trends and downturns over time. For mathematics, Sweden's achievement increased from 2011 to 2019, while Denmark and Finland experienced the opposite trend. In the case of Norway, their target grade changed from grade four to five in 2015, and hence their grade four achievements have been left out. Norway changed the target population of students from grades four and eight to grades five and nine to improve its comparability to other Nordic countries (Bergem et al., 2016; Kavli, 2018). Specifically, whereas Norwegian children start primary school at the age of six, Swedish, Danish, and Finnish children start primary school at the age of seven. As illustrated in Fig. 1.2, Norway's mathematics achievement in grade five decreased from 2015 to 2019.

In science, as depicted in Fig. 1.3, Denmark and Finland's achievements declined from 2011 to 2019. There were no significant changes for Norway between 2015 and 2019, while Sweden's achievement increased from 2011 to 2019. Sweden was

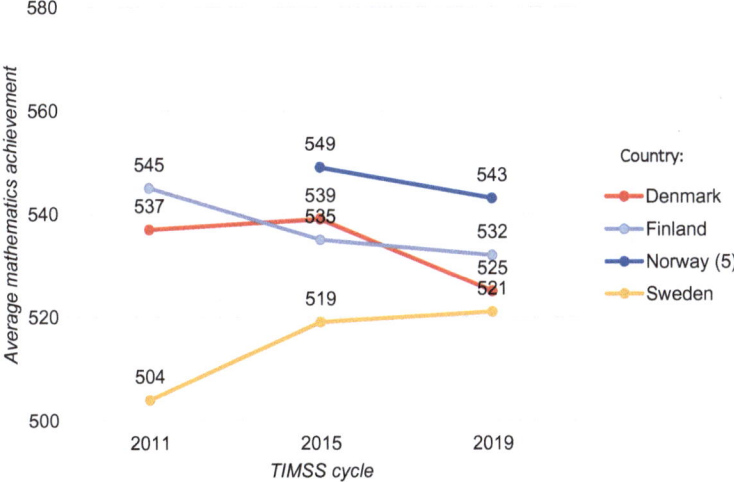

Fig. 1.2 Mathematics achievement in the Nordic countries over time

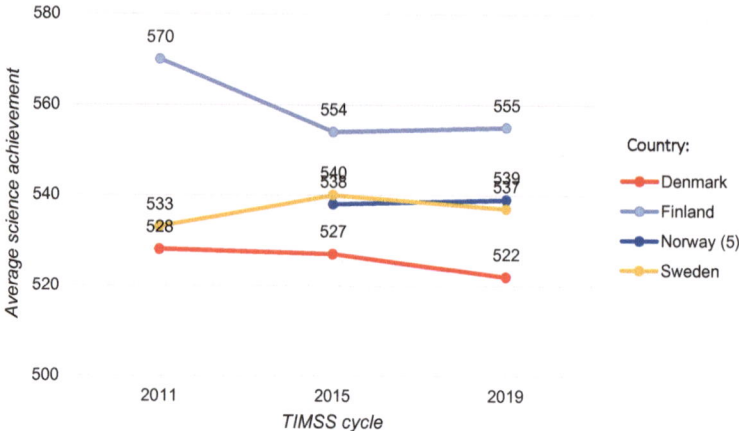

Fig. 1.3 Science achievement in the Nordic countries over time

the only country in this group to observe increased achievements in both science and mathematics from 2011 to 2019.

It is important to note that Iceland did not participate in TIMSS. They did participate in PISA. However, results from PISA are not comparable with results from TIMSS, as PISA measures mathematics and science literacy for 15-year-old students, while TIMSS measures students' competence in mathematics and science according to the participating countries' curricula in grade four (and eight).

To contextualize these changes in achievement, it is helpful to consider the standard deviation within the TIMSS scale. The scale was established in 1995 with a mean of 500 for participating countries and a standard deviation of 100 score points. The standard deviation is generally viewed as an indicator of educational equality and reflects the dispersion of students' achievement (Ferreira & Gignoux, 2014). Across these countries, a descriptive examination of the changes from 2011 to 2019 reveals an increased spread in student achievement in mathematics, as evidenced in Fig. 1.4. The results are more mixed in science (see Appendix 1 for all standard deviations and their standard errors in mathematics and science).

The increased standard deviation for the Nordic countries over time reflected in Fig. 1.4 warrants attention, as it suggests that teachers may be faced with heightened demands for differentiating instruction in classrooms where students' competencies exhibit greater variation than in the past.

Summary

The pattern across the Nordic countries points to declining achievements and an increased dispersion over time among students. This finding aligns with previous research highlighting the correlation between equity and achievement (e.g., Burroughs et al., 2019). It further points towards a negative trend in equity for the Nordic countries.

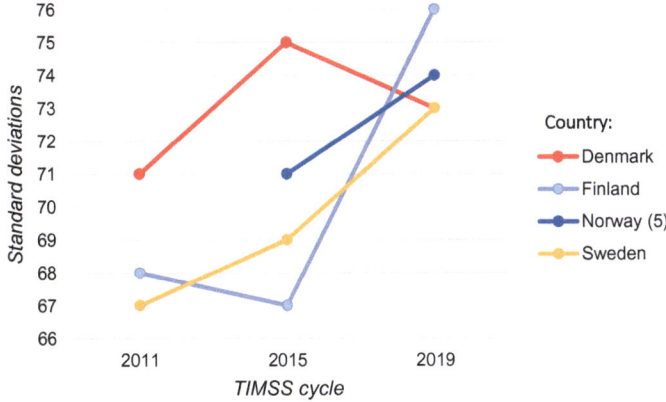

Fig. 1.4 Standard deviations in mathematics achievement in grade 4 (and grade 5 in Norway) in the Nordic countries over time

1.3.3 Results: Differences in Student Achievement Related to Language Spoken at Home

Whether the language spoken at home is the same as that of teachers' instruction can be regarded as a proxy for multiple dimensions of socioeconomic background and cultural capital, as posited within Pierre Bourdieu's theoretical framework (see e.g., Bourdieu & Wacquant, 2002). This framework suggests that the language spoken at home is not only indicative of SES and cultural capital but also has a more direct association with the overall process of language acquisition.

We first look at the proportions of students and the significant differences in the percentages of students who "sometimes" or "never" speak the language of the test at home over time and between countries. Note that the two categories "sometimes" and "never" were collapsed due to the small number of students in the "never" category.

Figure 1.5 presents the percentages of students in TIMSS who "sometimes" or "never" spoke the language of the test at home in the Nordic countries over time (for more detailed information, including standard errors, see Appendix 2).

The changes in the proportion of students predominantly using a language other than the national language at home present an intriguing phenomenon, particularly in the context of the most recent TIMSS cycle (2019). Our analysis reveals a statistically significant disparity in 2019 between Sweden on the one hand, and Norway and Finland on the other hand. For example, there were nine percentage points more Swedish students who "sometimes" or "never" speak the national language at home compared to Norway, and seven percentage points more compared to Finland. Upon examining the longitudinal changes within Sweden, a considerable divergence between 2015 and 2011 was observed. Overall, nearly one-fifth of all Swedish students fall under the category of students who "sometimes" or "never" speak the national language at home in all the last three TIMSS cycles. Conversely, Finland

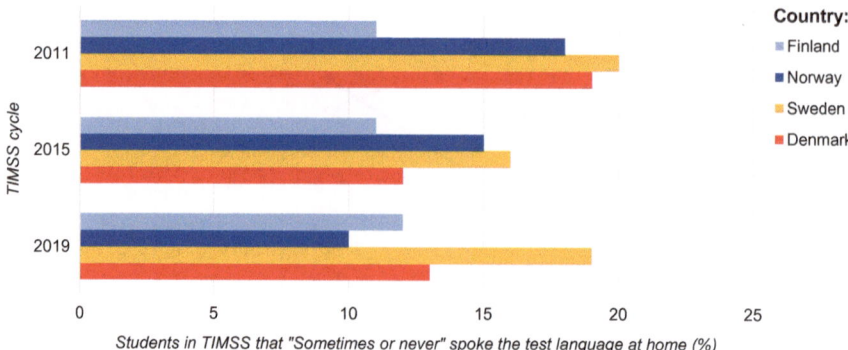

Fig. 1.5 The percentages of students in TIMSS that "sometimes" or "never" spoke the language of the test at home

has persistently maintained a lower proportion of students in this category among fourth-grade participants in TIMSS, as compared to other countries, and this pattern has remained stable across successive cycles.

Figure 1.6 presents the mean achievement gaps over time between students who "always" or "almost always" speak the language of the test at home and students who "sometimes" or "never" do. Substantial and statistically significant differences are found, with variations across cycles and subjects. The gap in science achievement is consistently higher than that of mathematics in all cycles and countries. One plausible explanation is that language accounts for more of the variance in science than mathematics because more advanced language skills are required in science.

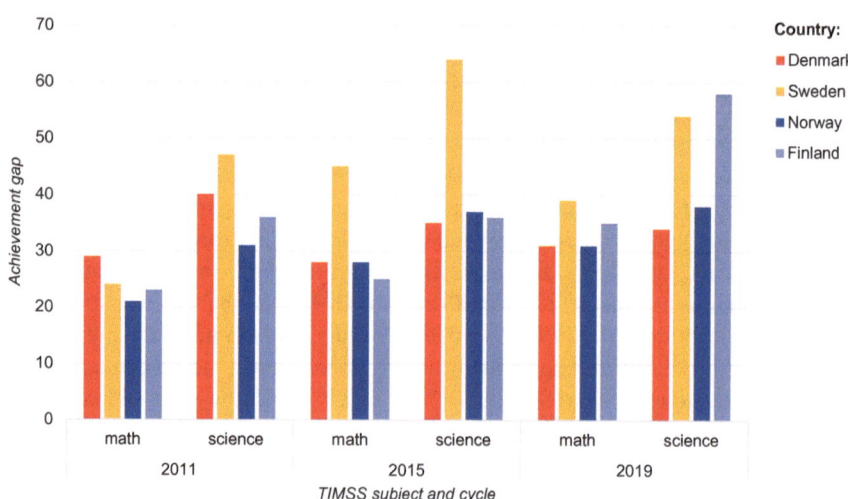

Fig. 1.6 Gap in the mean achievement between students who speak the language "sometimes" or "never" and "always" or "almost always" at home

There is a strong tendency for higher achievement gaps over time. This trend is evident from 2011 to 2019 across all countries and in both subjects, except for Denmark, where the achievement gap in science decreased over time.

The effect sizes of the gaps illustrated in Fig. 1.6, range from 0.30 percent in Norway in 2011 for mathematics to as high as 0.87 percent for science in Sweden in 2015 (see Appendix 3). A study by Hill et al. (2008), which analyzed annual mandatory assessments of American students, found an effect size 0.52 percent for the transition from third to fourth grade and 0.56 percent for the transition from fourth to fifth grade. These effect sizes are similar to the gaps illustrated in Fig. 1.6 and are presented in a table in Appendix 3, indicating that the achievement gaps reflect about one year of schooling between majority and minority language students (i.e., those who speak the language of the test "always or almost always" at home) and minority language students (those who speak the language of the test "sometimes" or "never" at home).

Summary

Section 1.3.3 investigated inequities associated with minority status, as indicated by students who "sometimes" or "never" speak the national language at home. The number of minority students varies across cycles and over countries, and the differences between the countries are especially striking. Similar to the results for the standard deviations, which indicated a tendency towards less equity over time for the Nordic countries, the findings for achievement gaps between majority and minority students also revealed a widening gap from 2011 to 2019, indicating a negative trend for equity.

1.3.4 Results: Variance in Achievements Explained by SES

The significance of socioeconomic background can be discerned through various approaches, pertaining not only to the conceptualization, operationalization, and quantification of socioeconomic background but also to the manner in which the computed estimations are correlated with student performance (Mittal et al., 2020). In large-scale assessment studies, such as TIMSS, the explained variance in a linear regression between achievement and one or multiple socioeconomic background variables is frequently utilized (e.g., Allerup et al., 2016; Mittal et al., 2020; Reimer et al., 2018; Strietholt & Strello, 2022).

In the TIMSS 2015 and 2019, a scale for home resources for learning has been developed, which offers insights into students' socioeconomic backgrounds. However, this scale has a large amount of missing data for some countries and is not available for Denmark for TIMSS 2011. Hence, when examining the changes across the three TIMSS cycles (2011, 2015, and 2019), we use the number of books in the students' home as this has previously been shown to be a valid and powerful proxy for student SES (Allerup et al., 2016; Gustafsson et al., 2011).

The concept of equal opportunity posits that the distribution of educational resources should be equitable, regardless of factors that ought to be inconsequential, such as gender, race, wealth, or geographical location. Within this context, both the gap analysis, which focuses on disparities across various groups, and the proportion of variance explained by student characteristics and home background (R^2) are found to be relevant indicators.

Figure 1.7 shows the proportion of variance in mathematics and science achievement explained by the number of books at home. A noticeable discrepancy can be observed in the variance explained by this variable across subjects; SES accounts for a greater proportion of the variation in student outcomes in science compared to mathematics. Furthermore, significant differences can be observed over time and across countries. At one end of the spectrum, the explained variance in students' mathematics scores in Finland in 2011 is a mere six percent in mathematics, while at the other end, Sweden demonstrates a considerably higher percentage of 14 percent in 2019 for mathematics.

Furthermore, Fig. 1.7 shows that a greater percentage of the variance in achievement is explained by the number of books at home in 2019 compared to 2011. This indicates a decrease in equity in the Nordic countries during this time period, and that students' home backgrounds matter more and more to their achievements.

Summary

This section examined the variance in student achievement explained by SES, as indicated by the number of books at home. The results point to less equity over time

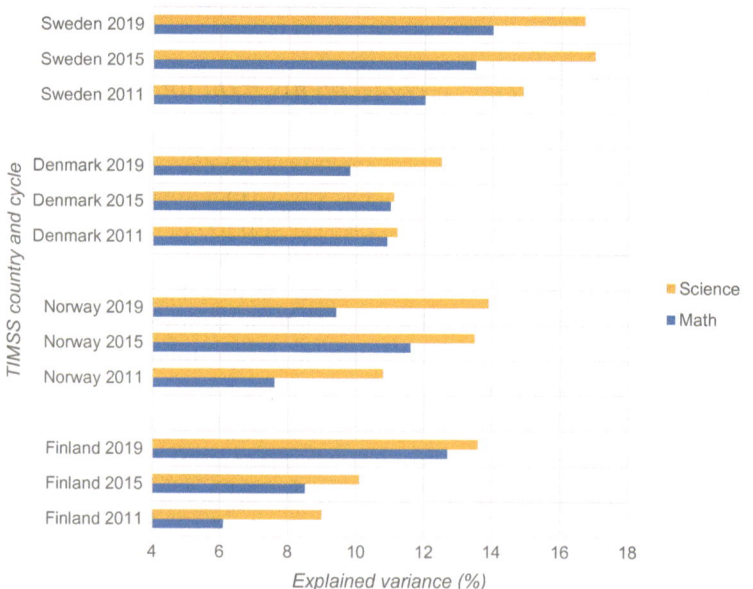

Fig. 1.7 Percent explained variance by the number of books at home in mathematics and science

for the Nordic countries since more variance is explained in 2019 than in 2011. The situation shows more pronounced differences in science compared to mathematics, and the results suggest a potential association between the achievement gap and the language spoken at home.

1.4 Conclusion

In this chapter, we have presented the rationale, aim, and structure of the book. Seeing how the main aim of the book is to investigate how teacher practices are related to achievement and equity over time, the chapter further provided the backdrop for the book in terms of achievements and inequalities in achievements for the Nordic countries. Various theoretical considerations concerning equity in education were discussed, supported by empirical insights at the national macro level. These empirical insights included student achievement over time, as well as three indicators of equity: the standard deviation of achievements as a measure of dispersion, language spoken at home as an indicator of minority, and number of books at home as an indicator of SES. All three indicators of equity suggest a negative development over time for the Nordic countries, along with a negative development of achievement.

The decreasing equity and achievements are in line with previous research that found correlations between the two (Burroughs et al., 2019; Kyriakides et al., 2018). Sweden, however, is an outlier and does not follow this pattern. Sweden was the only country with increased achievements in mathematics and science, while at the same time being the least equitable of the Nordic countries. One explanation could be that other factors, stronger than exogenous student characteristics, have promoted positive achievement trends in Sweden. Teachers are the heart and key of student learning, and Sweden has invested substantially in teacher education and professional development (Boesen, et al., 2015; Ringarp & Parding, 2018). In general, teachers have the potential to increase achievements and equity among students (e.g., Darling-Hammond, 2018; Praetorius et al., 2017). Yet, the disturbing picture emerging from our findings of negative trends for both achievements and equity in the Nordic countries, suggests that it is critical to examine *how* teachers may promote learning and equity in the Nordic countries. This is especially important to rectify the damages of the pandemic and counter further negative developments in these challenging times.

Appendices

Appendix 1 Standard Deviations with Standard Errors in Parentheses

	2019				2015				2011			
	Mathematics		Science		Mathematics		Science		Mathematics		Science	
Denmark	73	(1.1)	68	(1.1)	75	(1.6)	69	(1.3)	71	(2.0)	73	(1.9)
Finland	76	(1.4)	71	(1.5)	67	(1.2)	65	(1.7)	68	(1.5)	67	(1.5)
Norway	74	(1.4)	67	(1.4)	71	(1.4)	63	(1.5)	68	(1.9)	63	(1.3)
Sweden	73	(1.5)	74	(1.9)	69	(1.7)	73	(2.5)	67	(1.3)	75	(1.3)

Note Norway uses grade 4 as the target population in 2011. In 2015, Norway changed the target population to grade 5 to enhance comparability with other Nordic countries

Appendix 2 Percentages of Students Reporting on Language Spoken at Home in the TIMSS 2011, 2015, and 2019 Student Questionnaires for Mathematics

Country	Always or almost always	Sometimes or never
2019		
Denmark	87 (0.8)	13 (0.8)
Sweden	81 (1.4)	19 (1.4)
Norway	90 (0.7)	10 (0.7)
Finland	88 (1.0)	12 (1.0)
2015		
Denmark	88 (0.8)	12 (0.8)
Sweden	84 (1.3)	16 (1.3)
Norway	85 (1.2)	15 (1.2)
Finland	89 (0.7)	11 (0.7)
2011		
Denmark	81 (1.0)	19 (1.0)
Sweden	80 (1.0)	20 (1.0)
Norway	82 (1.1)	18 (1.1)
Finland	89 (0.7)	11 (0.7)

Note Responses for TIMSS have been recoded into a dichotomous variable with 1: "always or almost always" and "sometimes or never"

Appendix 3 Gap in Mean Achievement Between Students Who Speak the Language "Sometimes or Never" and "Always or Almost Always" at Home

Country	Mathematics			Science		
	Diff. in mean score	Cohen's d		Diff. in mean score	Cohen's d	
2019						
Denmark	31 (4.6)	0.42	***	34 (4.7)	0.50	***
Sweden	39 (5.3)	0.54	***	54 (5.1)	0.74	***
Norway	31 (4.6)	0.42	***	38 (4.4)	0.56	***
Finland	35 (5.3)	0.45	***	58 (4.4)	0.79	***
2015						

(continued)

(continued)

Country	Mathematics		Science	
	Diff. in mean score	Cohen's d	Diff. in mean score	Cohen's d
Denmark	28 (5.2)	0.37 ***	35 (4.9)	0.49 ***
Sweden	45 (5.8)	0.64 ***	64 (5.9)	0.87 ***
Norway	28 (5.9)	0.39 ***	37 (6.5)	0.58 ***
Finland	25 (6.2)	0.35 ***	36 (5.7)	0.52 ***
2011				
Denmark	29 (5.3)	0.41 ***	40 (6.0)	0.54 ***
Sweden	24 (4.0)	0.36 ***	47 (4.6)	0.63 ***
Norway	21 (4.0)	0.30 ***	31 (4.2)	0.49 ***
Finland	23 (5.6)	0.33 ***	36 (6.5)	0.52 ***

Note Responses for TIMSS have been recoded into a dichotomous variable with 1: "always or almost always" and "sometimes or never". *** denotes that the significance level $p < 0.001$

References

Allerup, P., Nøhr Belling, M., Kirkegaard, S. N., Thorndal Stafseth, V., & Torre, A. (2016). *Danske 4.-Klasseelever i TIMSS 2015: En international og national undersøgelse Af matematik og Natur/Teknologikompetence i 4. Klasse* [Danish 4th graders in TIMSS 2015: An international and national survey of mathematics and science/technology competence in 4th grade.]. Forlag1.dk.

Bergem, O. K., Kaarstein, H., & Nilsen, T. (2016). Vi kan lykkes i realfag—Resultater og analyser fra TIMSS 2015 [We can succeed in mathematics and science - results and analyses from TIMSS 2015]. Universitetsforlaget. https://doi-org/10.18261/97882150279999-2016

Boesen, J., Helenius, O., & Johansson, B. (2015). National-scale professional development in Sweden: Theory, policy, practice. *ZDM, 47*(1), 129–141. https://doi.org/10.1007/s11858-014-0653-4

Bourdieu, P., & Wacquant, L. (2002). *Refleksiv sociologi*. Hans Reitzels Forlag.

Burroughs, N., Gardner, J., Lee, Y., Guo, S., Touitou, I., Jansen, K. & Schmidt, W. (2019). Teacher effectiveness and educational equity. In: *Teaching for excellence and equity. IEA research for education*, vol 6. (pp. 101–136). Springer. https://doi.org/10.1007/978-3-030-16151-4_7

Darling-Hammond, L. (2018). From "separate but equal" to "No Child Left Behind": The collision of new standards and old inequalities. In *Thinking about schools* (pp. 419–437). Routledge.

Esping-Andersen, G. (1996). Welfare states in transition: National adaptations in global economies. *Sage*. https://doi.org/10.4135/9781446216941

Esping-Andersen, G. (2008). Childhood investments and skill formation. *International Tax and Public Finance, 15*, 19–44. https://doi.org/10.1007/s10797-007-9033-0

Espinoza, O. (2007). Solving the equity – Equality conceptual dilemma: A new model for analysis of the educational process. *Educational Research, 49*(4), 343–363. https://doi.org/10.1080/00131880701717198

Fauth, B., Atlay, C., Dumont, H., & Decristan, J. (2021). Does what you get depend on who you are with? Effects of student composition on teaching quality. *Learning and Instruction, 71*, 101355. https://doi.org/10.1016/j.learninstruc.2020.101355

Ferreira, F. H. G., & Gignoux, J. (2014). The measurement of educational inequality: Achievement and opportunity. *The World Bank Economic Review, 28*(2), 210–246. https://doi.org/10.1093/wber/lht004

Frønes, T. S., Pettersen, A., Radišić, J., Buchholtz, N. (2020). Equity, equality and diversity in the Nordic countries—Final thoughts and looking ahead. In *Equity, equality and diversity in the Nordic model of education*, (pp. 397–412). https://doi.org/10.1007/978-3-030-61648-9_16

Gustafsson, J. E., Yang Hansen, K., & Rosén, M. (2011). Effects of home background on student achievement in reading, mathematics, and science at the fourth grade. *TIMSS and PIRLS, 2011*, 181–287.

Kavli, A.-B. (2018). TIMSS and PISA in the Nordic countries. In Nordic Evaluation Network. In A. Wester, J. Välijärvi, J. K. Björnsson, & A. Macdonald (Eds.), *Northern lights on TIMSS and PISA* 2018 (pp. 11–30). The Nordic Council of Ministers.

Hansen, E. J. (1986). 4 inequality in the welfare state. *International Journal of Sociology, 16*(3–4), 107–121. https://doi.org/10.1080/15579336.1986.11769912

Yang Hansen, K., & Gustafsson, J. E. (2019). Identifying the key source of deteriorating educational equity in Sweden between 1998 and 2014. *International Journal of Educational Research, 93*, 79–90. https://doi.org/10.1016/j.ijer.2018.09.012

Hansen, E. J. (1973). The problem of equality in the Danish educational structure. *Acta Sociologica, 16*(4), 258–278. http://www.jstor.org/stable/4193960.

Hill, C. J., Bloom, H. S., Rebeck Black, A., & Lipsey, M. W. (2008). Empirical benchmarks for interpreting effect sizes in research. *Child Development Perspectives, 2*(3), 172–177. https://doi.org/10.1111/j.1750-8606.2008.00061.x

Kane, T. J., & Staiger, D. O. (2012). Gathering feedback for teaching: Combining high-quality observations with student surveys and achievement gains. Research Paper. MET Project. *Bill & Melinda Gates Foundation.* https://usprogram.gatesfoundation.org/news-and-insights/usp-resource-center/resources/gathering-feedback-on-teaching-combining-high-quality-observations-with-student-surveys-and-achievement-gains--report

Kyriakides, L., Creemers, B., & Charalambous, E. (2018). Investigating the relationship between quality and equity: Secondary analyses of national and international studies. *Equity and Quality Dimensions in Educational Effectiveness*, 97–125. https://doi.org/10.1007/978-3-319-72066-1_5

Martin, M. O., von Davier, M., & Mullis, I. V. S. (Eds.). (2020). *Methods and procedures: TIMSS 2019 technical report.* TIMSS & PIRLS International Study Center, Boston College. https://timssandpirls.bc.edu/timss2019/methods

Mittal, O., Nilsen, T., & Björnsson, J. K. (2020). Measuring equity across the Nordic education systems—Conceptual and methodological choices as implications for educational policies. In *Equity, equality and diversity in the Nordic model of education*, (pp. 43–71). Springer. https://doi.org/10.1007/978-3-030-61648-9_3

Mullis, I. V., Martin, M. O., Foy, P., Kelly, D. L., & Fishbein, B. (2020). *TIMSS 2019 international results in mathematics and science.* TIMSS & PIRLS International Study Center Boston College. https://timssandpirls.bc.edu/timss2019/international-results

Nilsen, T., Björnsson, J. K., & Olsen, R. V. (2018). Hvordan har likeverd i norsk skole endret seg de siste 20 årene? [How has equality in Norwegian schools changed in the last 20 years?] In J. K. Björnsson, & R. V. Olsen, (Red.) *Tjue år med TIMSS og PISA i Norge: Trender og nye analyser* [Twenty years of TIMSS and PISA in Norway: Trends and new analyses] (pp. 150–172). https://doi.org/10.18261/9788215030067-2018-08

OECD. (2012). Equity and quality in education: Supporting disadvantaged students and schools. *OECD Publishing Paris.* https://doi.org/10.1787/9789264130852-en

OECD. (2019), *PISA 2018 results (Volume II): Where all students can succeed.* PISA. OECD Publishing. https://doi.org/10.1787/b5fd1b8f-en

Oftedal Telhaug, A., Asbjørn Medias, O., & Aasen, P. (2006). The Nordic model in education: Education as part of the political system in the last 50 years. *Scandinavian Journal of Educational Research, 50*(3), 245–283. https://doi.org/10.1080/00313830600743274

Pedersen, H. (2014). Education policy making for social change: A post-humanist intervention. *Policy Futures in Education, 8*(6), 683–696. https://doi.org/10.2304/pfie.2010.8.6.682

Praetorius, A.-K., Lauermann, F., Klassen, R. M., Dickhäuser, O., Janke, S., & Dresel, M. (2017). Longitudinal relations between teaching-related motivations and student-reported teaching quality. *Teaching and Teacher Education, 65*, 241–254. https://doi.org/10.1016/j.tate.2017.03.023

Reimer, D., Skovgaard Jensen, S., & Christrup Kjeldsen, C. (2018). Social inequality in student performance in the Nordic countries: A comparison of methodological approaches. In *Northern lights on TIMSS and PISA 2018*, (pp. 31–59). TemaNord 524. Nordic Council of Ministers. https://doi.org/10.6027/TN2018-524

Ringarp, J., & Parding, K. (2018). I otakt med tiden?: Lärarprofessionens ställning sett via lärarut-bildningens utveckling i Sverige, 1962–2015 [At odds with time? The position of the teaching profession seen through the development of teacher education in Sweden]. In M. Buchardt, J. E. Larsen, & K. B. Staffensen (Eds.), *Professionerne og deres uddannelser* [The professionals and their education] (pp. 49–68). Köpenhamn, Denmark: Selskabet for skole- og uddannelseshistorie.

Schmidt, W. H., Houang, R. T., Cogan, L., Blömeke, S., Tatto, M. T., Hsieh, F. J., Santillan, M., Bankov, K., Han, S. I., & Cedillo, T. (2008). Opportunity to learn in the preparation of mathematics teachers: Its structure and how it varies across six countries. *ZDM, 40*(5), 735–747. https://doi.org/10.1007/s11858-008-0115-y

Sen, A. (1980). *Equality of what? The Tanner Lecture on Human Values 1197–220*. Cambridge University Press.

Sen, A. (2008). The idea of justice1. *Journal of Human Development, 9*(3), 331–342. https://doi.org/10.1080/14649880802236540

Strietholt, R., & Strello, A. (2022). Socioeconomic inequality in achievement. In T. Nilsen, A. Stancel-Piątak, & J. E. Gustafsson (Eds.), *International handbook of comparative large-scale studies in education: Perspectives, methods and findings* (p. 201). Springer International Handbooks of Education. https://doi.org/10.1007/978-3-030-88178-8_11

Tröhler, D., Hormann, B., Tveit, S., & Bostad, I. (2022). The Nordic education model in context: Historical developments and current renegotiations. *Taylor & Francis*. https://doi.org/10.4324/9781003218180

UNESCO Institute for Statistics. (2018). *Handbook on measuring equity in education*. UNESCO Institute for Statistics.

UNESCO. (2015). *EFA global monitoring report. Education for all 2000–2015: Achievements and challenges*. UNESCO. https://doi.org/10.54676/LBSF6974

Wahlström, N. (2022). Equity, teaching practice and the curriculum. *Routledge*. https://doi.org/10.4324/9781003218067

Christian Christrup Kjeldsen Ph.D., serves as an Associate Professor at the Danish School of Education (DPU) Aarhus University. He is the Centre Director for the National Centre for School Research (NCS) and the lead authority for all IEA (International Association for the Evaluation of Educational Achievement) projects. Dr. Kjeldsen specializes in empirical research, utilizing statistical methods to explore the relationship between school achievement and various forms of social background and status. He currently holds the position of National Research Coordinator for the IEA's TIMSS (Trends in International Mathematics and Science Study).

Trude Nilsen Trude Nilsen is a research professor at the University of Oslo. She is a leader of Strand 1 at CREATE—Centre for Research on Equality in Education, a leader of the research group LEA, and of funded research projects. She has been engaged as an international external expert for IEA's TIMSS and for OECD's TALIS studies. Her research focuses on teaching quality, educational equality, school climate, and applied methodology including causal inferences.

Jenna Hiltunen has an MSc degree in mathematics and specializes in mathematics education. Hiltunen has been working as a project researcher at the Finnish Institute for Educational Research mainly in TIMSS and PISA studies since 2014, especially responsible for mathematics content. Alongside research in large-scale assessments, she is preparing for a Ph.D. about dialogic argumentation in mathematics.

Nani Teig is an associate professor at the University of Oslo, Norway. Her research focuses on inquiry-based teaching, scientific reasoning, teaching quality, and academic resilience. She integrates multilevel analyses using data from videos, surveys, assessments, and computer log files. Dr. Teig has received several awards and fellowships, including the Global Education Award, Bruce H. Choppin Dissertation Award, Young CAS Fellow, and UNESCO GEM Fellow.

Chapter 2
Theoretical Framework of Teacher Practice

Nani Teig⊙, Trude Nilsen⊙, and Kajsa Yang Hansen⊙

2.1 Introduction

Understanding the factors that contribute to effective and equitable teacher practice is of top priority within educational research. Researchers strive to determine how various aspects of teacher practice can be customized to provide optimal learning opportunities for diverse student populations (Dudek et al., 2019; Wallace, 2009). As the world becomes increasingly interconnected and diverse, it is imperative for educational systems to adapt and respond to the varying needs of students from different cultural, linguistic, and socioeconomic backgrounds. By gaining novel insights into effective and equitable teacher practice, we can establish a solid foundation for evidence-based professional practice and teacher education that aims to enhance student outcomes and narrow the gap in educational disparities.

N. Teig (✉)
Department of Teacher Education and School Research, University of Oslo, Postboks 1099 Blindern, 0317 Oslo, Norway
e-mail: nani.teig@ils.uio.no

T. Nilsen
Department of Teacher Education and School Research, CREATE—Centre for Research on Equality in Education (project number 331640), University of Oslo, Postboks 1099, Blindern, 0317 Oslo, Norway
e-mail: trude.nilsen@ils.uio.no

K. Yang Hansen
Department of Education and Special Education, University of Gothenburg, P.O. Box 300, 40530 Gothenburg, Sweden
e-mail: kajsa.yang-hansen@ped.gu.se

© The Author(s) 2024
N. Teig et al. (eds.), *Effective and Equitable Teacher Practice in Mathematics and Science Education*, IEA Research for Education 14,
https://doi.org/10.1007/978-3-031-49580-9_2

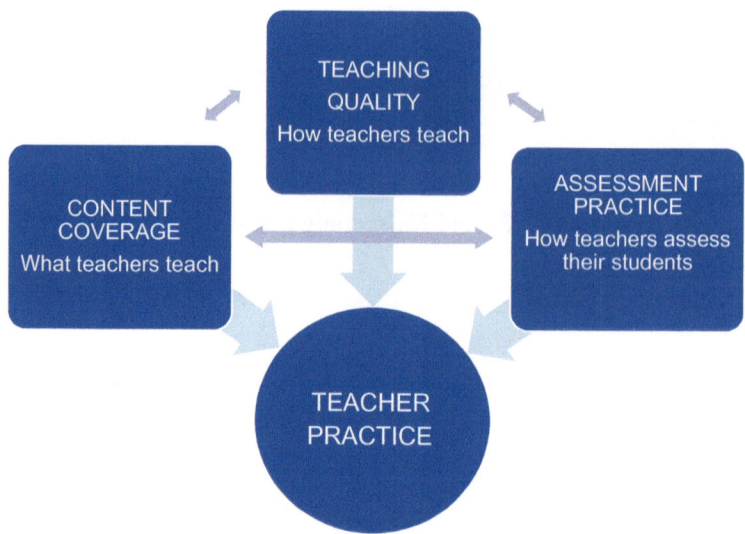

Fig. 2.1 The conceptual framework of teacher practice

But what exactly is teacher practice? The concept of "teacher practice" encompasses a wide range of teachers' work and responsibilities within educational settings (Klein, 2012). Teacher practice may relate to instructional methods and strategies, classroom assessments, lesson planning, or curriculum implementation and can be influenced by teachers' beliefs and attitudes about the nature of teaching and learning (Anderman & Anderman, 2020; Denessen et al., 2022; Wallace, 2009). This book specifically focuses on three aspects of teacher practice that directly impact student learning. As depicted in Fig. 2.1, the *content coverage* reveals what teachers teach, *teaching quality* describes how teachers deliver the content, and *assessment practice* shows how teachers assess their students' learning outcomes.

These three aspects of teacher practice are interconnected and play a significant role in shaping students' learning experiences and their subsequent academic outcomes (Fauth et al., 2014; Panadero et al., 2017; Schmidt et al., 2021). This chapter provides an in-depth exploration of these aspects and discusses their interrelationships, with specific emphasis on mathematics and science learning.

2.2 What Teachers Teach: Content Coverage

Within the context of teacher practice, what teachers teach, or content coverage serves as a foundation for learning, determining the scope and depth of students' learning, influencing their understanding, critical thinking skills, and overall academic growth. Content coverage refers to the amount of material that is covered or taught in a particular subject, making it an essential aspect of education as it ensures students to have a fundamental understanding of the subject matter (Porter, 2002). Content coverage can vary depending on the educational level, subject matter, and objectives of the course. Sufficient content coverage can provide students exposure to all the necessary topics, concepts, and skills outlined in the curriculum or educational standards. Conversely, inadequate content coverage can limit students' opportunities to learn, potentially leading to knowledge gaps and hindering their overall academic achievement.

Content coverage represents a critical aspect of any curriculum, as it outlines the subject matter students will be exposed to and the knowledge they are expected to acquire. The relationship between content coverage and curriculum is vital, as it ensures the fulfilment of educational goals and objectives stipulated within the curriculum. It is widely observed that students typically perform better on topics they have been taught, compared to those they have not.

Content coverage has also been conceptualized as "opportunity to learn" (OTL) in large-scale studies conducted by the International Association for the Evaluation of Educational Achievement (IEA) including the Trends in International Mathematics and Science Study (TIMSS) (see e.g., Schmidt et al., 1997). Generally, OTL encapsulates more than just content coverage; it refers to the extent to which students have access to quality learning experiences, time, and resources that support the acquisition of knowledge and skills (Floden, 2002; Perry et al., 2023; Schmidt et al., 2021). Content coverage is an essential component of OTL, as it determines what students are exposed to during their time in the classroom. There is a strong relationship between content coverage and OTL. In order for students to comprehend the material being covered, it is imperative that they are provided with sufficient OTL. Inadequate time or resources can impede students' ability to fully grasp the material, ultimately resulting in suboptimal academic performance. Similarly, when the scope of the content coverage is too extensive or intricate, students may not have enough time to fully understand the subject matter, which can also lead to poor academic performance.

In the TIMSS framework, content coverage is distinguished into three key components (Mullis & Martin, 2017). The *intended curriculum*, prescribed at the system level, refers to the officially prescribed learning objectives, standards, and subject matter that students are expected to learn, as outlined by educational authorities. The *implemented curriculum*, manifested at the classroom level, is the actual content delivered by teachers in the classroom, which may differ from the intended curriculum due to factors such as teachers' competence, school or classroom resources, and students' backgrounds. The *attained curriculum* refers to the knowledge, skills, and competencies that students acquired as a result of their educational experiences. Better alignment between educational goals (the intended and implemented curricula) with educational outcomes (the attained curriculum) is an important characteristic of effective teacher practice (Daus et al., 2018).

In general, when compared to other aspects of teacher practice, such as teaching quality and assessment practice, the extent of content coverage largely depends on the intended curriculum at the national level. However, Nordic countries use multi-year curricula, which span across several years or grades and outline the learning objectives, topics, and skills that students are expected to acquire over that period (see Chap. 4). For instance, in Norway, the curriculum is organized into three-year cycles, with the first cycle covering grades 1–4, the second cycle covering grades 5–7, and the third cycle covering grades 8–10. Unlike annual curricula that provide more specific guidance on what should be covered in a particular grade level, multi-year curricula give teachers a certain degree of autonomy and flexibility to decide when and how to cover specific topics and learning objectives within the curriculum cycle. Consequently, content coverage emerges as a key aspect of teacher practice, as teachers are tasked with selecting suitable topics and adjusting their instruction and assessment strategies to meet their students' needs and interests. At the same time, they are responsible for ensuring the required curriculum is covered and providing a coherent learning experience for students over several years.

This book conceptualizes content coverage as student exposure to TIMSS' mathematics and science topics in grades four and five. Content coverage, in this context, refers to the coverage of topics in the three content domains of mathematics (number, geometry, and data) and of science (life science, physical science, and earth science). Teachers reported whether and when they have covered the topics. This conceptualization is applied in Chaps. 4, 6, and 8. Chapter 4 further examines the alignment between content coverage (implemented curricula) with educational goals (intended curricula) and educational outcomes (attained curricula). Meanwhile, Chaps. 6 and 8 investigate the relations between content coverage and student achievement across the various content domains of mathematics and science.

2.3 How Teachers Teach: Teaching Quality

Teaching quality is a multifaceted construct that has garnered significant attention in the field of education due to its pivotal role in shaping student learning outcomes and experiences. Various definitions of teaching quality have emerged in the literature, reflecting its diverse aspects and the complexity of the teaching process (Senden et al., 2022). Some scholars interpret teaching quality through the lens of generic, domain-specific aspects, or a blend of both (Blömeke et al., 2016). Others approach it through specific instructional practices, such as differentiated instruction, problem-based learning, inquiry-based teaching, and formative assessment (Hattie, 2009; Ko & Sammons, 2013; Muijs & Reynolds, 2017). This perspective underscores the importance of adapting teacher instruction with the diverse learning needs and styles of students in order to maximize their potential for success. Additionally, the concept of teaching quality has been closely linked to teacher effectiveness and the ability to create a supportive and engaging learning environment (Goe et al., 2008). Another perspective emphasizes the necessity of ongoing professional development and the capacity to adapt teaching practices in response to students' needs and the dynamic nature of educational contexts (Darling-Hammond et al., 2017).

Despite the varying perspectives on teaching quality, a consistent feature in the literature is the recognition that teaching quality serves as a crucial determinant of student achievement, motivation, and overall educational success. This book adopted

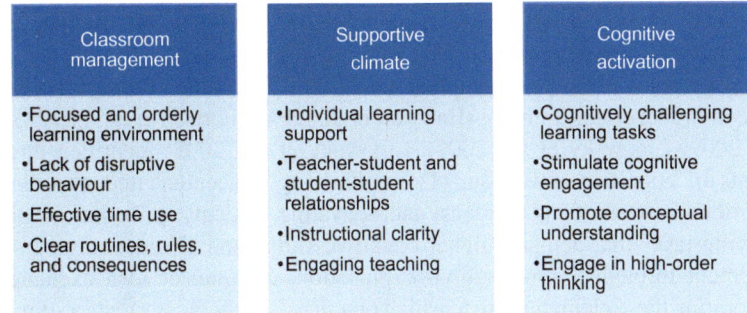

Fig. 2.2 Various aspects of teaching quality. *Note* Figure adopted from Klieme et al. (2009)

the three basic dimensions of teaching quality from Klieme et al. (2009) as a theoretical framework, which encompasses classroom management, supportive climate, and cognitive activation. (Fig. 2.2).

Classroom management is a fundamental and the most generic aspect of teaching quality. It refers to the strategies, techniques, and processes explored by teachers to establish and maintain a well-organized, focused, and orderly learning environment (Praetorius et al., 2018). An orderly classroom environment with minimal disruptions allows students to focus on learning and making the most of their educational experience (Freiberg et al., 2020; Marder et al., 2023). It involves effective time management, task-oriented student behavior, consistent enforcement of rules and consequences, and the establishment of routines.

The positive impact of classroom management on student outcomes across various subjects is more robust than other aspects of instructional quality (Korpershoek et al., 2016; Senden et al., 2023). High-achieving classrooms often exhibit effective classroom management, which fosters a conducive learning atmosphere and encourages student engagement (Dijk et al., 2019; Korpershoek et al., 2016).

Supportive climate refers to the overall classroom environment that facilitates positive student learning experiences, including teacher support, classroom interaction (teacher-student and student–student relationships), and instructional clarity. Creating a supportive climate in a classroom entails the teacher's ability to foster an atmosphere that promotes students' intellectual, social, and emotional development. This involves providing personalized support to address the individual students' unique needs, establishing clear expectations, and utilizing varied instructional approaches to enhance understanding (Senden et al., 2022).

The need for a supportive climate in mathematics and science classrooms is particularly critical due to the complex and abstract nature of the subjects. Students often encounter challenging concepts and problem-solving tasks in these areas. Consequently, establishing a safe and supportive environment, where students feel comfortable to ask questions and seek clarification, is crucial. This nurturing environment not only fosters student engagement and motivation but also nurtures their interest, curiosity, and enthusiasm for these subjects (Teig & Nilsen, 2022).

Cognitive activation represents a domain-specific aspect of teaching quality, involving instructional approaches and learning tasks that stimulate students' cognitive engagement, promote conceptual understanding, and encourage students to engage in higher-order thinking (Baumert et al., 2010; Förtsch et al., 2017; Klieme et al., 2009; Lipowsky et al., 2009). Lipowsky et al. (2009) identified three key elements of cognitive activation: (1) emphasizing conceptual understanding and connections between facts or ideas, and activating students' prior knowledge, (2) employing tasks that demand higher cognitive skills; and (3) encouraging student engagement through argumentation, explanation, critique, or idea exchange. By incorporating these elements into teaching practices, teachers can create a stimulating learning environment that enhances students' critical thinking and problem-solving skills.

The level of cognitive activation largely depends on the selection and implementation of tasks and activities in the classrooms (Baumert et al., 2010; Lipowsky et al., 2009). Cognitive activation is more likely to occur when teachers present challenging tasks that stimulate students' thinking, encourage them to recognize connections between new content and their existing knowledge, and promote discussions about potential problem solutions. Additionally, exploring multiple approaches to solve a problem and emphasizing the importance of self-reflection can further enhance cognitive activation (Baumert et al., 2010; Lipowsky et al., 2009). On the other hand, cognitive activation is less likely to occur if teachers merely view learning as the one-way transmission of subject knowledge (Lipowsky et al., 2009).

Cognitive activation can be distinguished into general and subject-specific forms (Schlesinger et al., 2018; Teig et al., 2019). General cognitive activation represents practices applicable across all classrooms, regardless of the subject domain. In contrast, subject-specific cognitive activation relates to the unique aspects of cognitive activation that typically characterize a particular subject domain. Cognitive activation may involve students independently applying what they have learned to new problem situations, linking content with their everyday lives, and expressing their ideas or explaining their answers to challenging exercises. Typical examples of cognitive activation in mathematics include providing students with the opportunity to deal with mathematical proof and engage in other mathematical processes, including problem-solving, modeling, or reasoning (Schlesinger et al., 2018; Sigurjónsson, 2023). In science classrooms, cognitive activation typically involves students in scientific inquiry practices, such as formulating research questions, designing and conducting investigations, and analyzing and interpreting data (Teig et al., 2019; Teig et al., 2022). Inquiry-based cognitive activation strategies enable students to learn about scientific content and explore the nature of science more deeply through first-hand experience in scientific investigations (Teig et al., 2019). Both general and subject-specific cognitive activation play vital roles in determining the quality of teaching.

The empirical chapters in this book examine various dimensions of teaching quality. Chapter 5 explores the trends in classroom management, supportive climate (specifically on teacher support and instructional clarity), and cognitive activation as well as their relations to mathematics and science achievement. Chapter 7 investigates whether changes in supportive climate and cognitive activation are related to the changes in achievement in both subjects. Meanwhile, Chapter 9 delves into similar dimensions by focusing on their roles in mitigating socioeconomic and ethnic disparities in mathematics.

2.4 How Teachers Assess Their Students: Assessment Practice

Teacher assessment practice encompasses a range of methods and strategies used by teachers to gather evidence of their students' understanding (Black & Wiliam, 1998; Popham, 1999). This evidence serves as a basis for important educational decisions, including adapting instruction, selecting assignments, providing feedback, assigning grades, and planning lessons (Black & Wiliam, 1998; Gardner et al., 2010; Herppich et al., 2018; Popham, 1999).

In general, three main types of assessment can be identified: assessment *for* learning, assessment *of* learning, and assessment *as* learning, each serving distinctive objectives and functions in educational settings. *Assessment for learning*, also known as formative assessment, is used to inform and improve the teaching and learning process (Black & Wiliam, 1998; Schildkamp et al., 2020). It takes place during the learning process and provides teachers with valuable information about students' understanding, progress, and misconceptions (Schildkamp et al., 2020). Teachers can then use this information to adjust their instruction, provide feedback, and address any learning gaps. *Assessment of learning* or summative assessment is typically conducted at the end of a unit, course, or academic year (Gao et al., 2020; Harlen, 2007). Its primary purpose is to evaluate students' overall achievement and mastery of specific learning objectives (Harlen, 2007), for example, through standardized tests, final exams, and end-of-term projects. *Assessment as learning* emphasizes the students' active involvement in their own learning process (Panadero et al., 2017; Popham, 1999). It promotes metacognition, self-assessment, and reflection, enabling students to become more independent and self-regulated learners (Panadero et al., 2017).

To accommodate these various assessment types, educators utilize a wide range of assessment practices, from traditional exams and quizzes to more innovative approaches like project-based assessments, peer evaluations, and learning journals or reflection logs. One common example of assessment practice is the assignment of homework, which among other things allows teachers to gauge students' understanding of the material, helps students practice and reinforce skills learned in the classroom, and can also serve as a way for students to learn new content (Fernández-Alonso & Muñiz, 2022). Effective assessment practice enables teachers to identify strengths and weaknesses and efficacy of their teaching methods and allows them to adjust their instruction to better meet the needs of their students (Black &

Wiliam, 1998; Popham, 1999). Furthermore, effective assessment practice provides valuable feedback to students and offers guidance for improvement, helping them understand their progress (Gardner et al., 2010). Assessment practices can foster a positive collaboration between teachers and students, ultimately contributing to the development of student outcomes (Black & Wiliam, 1998; Muijs & Reynolds, 2017).

Assessment practices can also be used to establish high standards and expectations for all students, including those from socioeconomically disadvantaged backgrounds (Andrade & Brookhart, 2020; Panadero et al., 2017). The practices enable teachers to identify learning gaps among disadvantaged students, thus allowing them to tailor instruction, resources, and support accordingly. By upholding rigorous standards for every student, teachers can promote a culture of achievement and ensure equal opportunity to succeed.

In mathematics and science classrooms, assessment practices hold a significant place due to the complex and abstract nature of these subjects (Gao et al., 2020). Mathematics and science often require higher-order thinking, problem-solving, and critical analysis, making it essential for teachers to employ adequate assessment strategies to ensure positive learning outcomes and experiences for students. Additionally, assessment practices in mathematics and science may promote metacognition, persistence, and resilience, as students are encouraged to reflect on their learning processes and work through challenges (Gao et al., 2020). By using assessment data to inform instruction, teachers can ensure that their students are developing a deep understanding of the material and are able to apply their knowledge to real-world situations (Black & Wiliam, 1998; Muijs & Reynolds, 2017).

Subsequent chapters in this book delve into various aspects of assessment practice in mathematics and science. Chapter 5 explores the trends in homework frequency, homework time, how teachers use homework in the classroom (referred to as in-class homework discussion), and the emphasis teachers place on assessment strategies. Chapter 7 further investigates whether changes in homework frequency, time spent on homework, and in-class homework discussion correspond to the changes in achievement in both subjects. Meanwhile, Chapter 9 scrutinizes teachers' emphasis on assessment strategies in mitigating disparities in mathematics achievement. Together, these chapters provide a comprehensive examination of assessment practices and their implications for student learning outcomes.

2.5   An Integrated Framework

To emphasize the importance of aligning curriculum objectives with assessment measures and understand the interconnection between content coverage, teaching quality, assessment practice, and students' learning outcomes, the book adopts the theoretical model of Potential Educational Experiences (Schmidt et al., 1997). This model describes the dynamic mechanism between content coverage, teaching quality, and assessment practices in facilitating effective learning experiences for all students (Floden, 2002; Perry et al., 2023; Schmidt et al., 2021).

In the Potential Educational Experiences model, the implemented curriculum functions as a mediator between the intended curriculum and the curriculum carried out. It serves as a representation of the desired learning experiences outlined in the intended curriculum, and at the classroom level, it can be referred to as the "opportunity" provided to students. The choices made by schools and teachers, such as student grouping, timetable structuring, and resource selection, all have implications for the educational opportunities available. The model also highlights direct and indirect effects on the attained curriculum, considering various antecedents and contexts at the system level, school and classroom levels, and student level. These antecedents may include, for example, teacher characteristics, teaching practice, learning conditions, and student attributes (see Fig. 2.3).

The model identifies three main channels through which the implemented curriculum impacts the attained curriculum. These channels involve the influence of student characteristics and peers on teaching quality, the effects of teacher practice-related factors (e.g., content coverage, instructional activities, and supportive functions) on student achievement, and the impact of organizational differentiation on teacher resources and teaching support. In this book, our primary examination focuses on the second channel of teacher practice and its impact on students' learning outcomes. Specifically, we delve into content coverage, teaching quality, and assessment practice, either independently to analyze changes over time, or in an integrated manner to explore the interrelationship between these constructs.

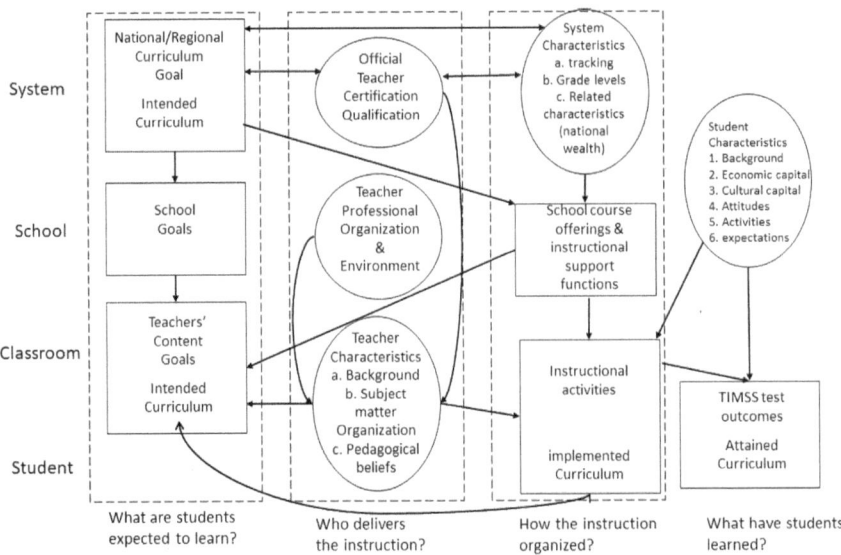

Fig. 2.3 The TIMSS model of potential educational experiences. *Source* Schmidt et al. (1997, p. 188)

2.6 Closing Remarks

One of the primary challenges in investigating effective and equitable teacher practice involves acknowledging that different aspects of teaching practice—what teachers teach (content coverage), how teachers teach (teaching quality), and how teachers assess their students (assessment practice)—are not static or isolated factors. Instead, they are dynamic and interconnected factors that exert their influence on student outcomes, as depicted in Fig. 2.3. Adequate content coverage ensures students' opportunity to learn essential concepts and competencies. High-quality teaching fosters a positive learning environment, encourages student engagement with the subject matter, and promotes deep understanding. Effective assessment practices help teachers to identify student strengths and weaknesses, allowing for tailored instruction and support, which in turn contribute to improved student outcomes. When these three aspects of teacher practice are well-aligned and consistently applied, they collectively create a cohesive and effective learning experience for students, ultimately leading to better outcomes.

Understanding the impact of any single aspect of teacher practice requires considering its relationship with other aspects. For instance, content coverage and teaching quality are inextricably linked. A solid understanding of the curriculum empowers teachers to determine, integrate, and present mathematics and science content in a coherent and meaningful way, making it more accessible and engaging for students. Similarly, teacher assessment practices play a crucial role in shaping both content coverage and teaching quality, as they provide valuable feedback on student learning and progress that can inform instructional decisions and adaptations. These interconnected aspects of teacher practice work together to create a conducive environment that fosters academic growth and success for all students. By recognizing the interdependence of these aspects, researchers and practitioners can develop a more comprehensive understanding of effective and equitable teacher practice, essential for improving educational outcomes for all students.

References

Anderman, E. M., & Anderman, L. H. (2020). *Classroom motivation: Linking research to teacher practice*. Routledge.

Andrade, H. L., & Brookhart, S. M. (2020). Classroom assessment as the co-regulation of learning. *Assessment in Education: Principles, Policy & Practice, 27*(4), 350–372. https://doi.org/10.1080/0969594x.2019.1571992

Baumert, J., Kunter, M., Blum, W., Brunner, M., Voss, T., Jordan, A., Klusmann, U., Krauss, S., Neubrand, M., & Tsai, Y.-M. (2010). Teachers' mathematical knowledge, cognitive activation in the classroom, and student progress. *American Educational Research Journal, 47*(1), 133–180. https://doi.org/10.3102/0002831209345157

Black, P., & Wiliam, D. (1998). *Inside the black box: Raising standards through classroom assessment*. Granada Learning.

Blömeke, S., Olsen, R. V., & Suhl, U. (2016). Relation of student achievement to the quality of their teachers and instructional quality. In: T. Nilsen & J. E. Gustafsson (Eds.), Teacher quality, instructional quality and student outcomes. IEA Research for Education, vol 2. Springer. https://doi.org/10.1007/978-3-319-41252-8_2

Darling-Hammond, L., Hyler, M. E., & Gardner, M. (2017). *Effective teacher professional development*. Learning Policy Institute.

Daus, S., Nilsen, T., & Braeken, J. (2018). Exploring content knowledge: Country profile of science strengths and weaknesses in TIMSS. Possible implications for educational professionals and science research. *Scandinavian Journal of Educational Research, 63*(7), 1102–1120. https://doi.org/10.1080/00313831.2018.1478882

Denessen, E., Hornstra, L., van den Bergh, L., & Bijlstra, G. (2022). Implicit measures of teachers' attitudes and stereotypes, and their effects on teacher practice and student outcomes: A review. *Learning and Instruction, 78*, 101437. https://doi.org/10.1016/j.learninstruc.2020.101437

Dijk, W., Gage, N. A., & Grasley-Boy, N. (2019). The relation between classroom management and mathematics achievement: A multilevel structural equation model. *Psychology in the Schools, 56*(7), 1173–1186. https://doi.org/10.1002/pits.22254

Dudek, C. M., Reddy, L. A., & Lekwa, A. (2019). Measuring teacher practices to inform student achievement in high poverty schools: A predictive validity study. *Contemporary School Psychology, 23*, 290–303. https://doi.org/10.1007/s40688-018-0196-8

Fauth, B., Decristan, J., Rieser, S., Klieme, E., & Büttner, G. (2014). Student ratings of teaching quality in primary school: Dimensions and prediction of student outcomes. *Learning and Instruction, 29*, 1–9. https://doi.org/10.1016/j.learninstruc.2013.07.001

Fernández-Alonso, R., & Muñiz, J. (2022). Homework: Facts and fiction. In: T. Nilsen, A. Stancel-Piątak, & J. E. Gustafsson (Eds.), *International handbook of comparative large-scale studies in education*. Springer https://doi.org/10.1007/978-3-030-88178-8_40

Floden, R.E. 2002. The measurement of opportunity to learn. In National Research Council (Ed.), *Methodological advances in cross-national surveys of education achievement* (pp. 231–266). National Academies Press.

Förtsch, C., Werner, S., Dorfner, T., von Kotzebue, L., & Neuhaus, B. J. (2017). Effects of cognitive activation in biology lessons on students' situational interest and achievement. *Research in Science Education, 47*(3), 559–578. https://doi.org/10.1007/s11165-016-9517-y

Freiberg, H. J., Oviatt, D., & Naveira, E. (2020). Classroom management meta-review continuation of research-based programs for preventing and solving discipline problems. *Journal of Education for Students Placed at Risk (JESPAR), 25*(4), 319–337. https://doi.org/10.1080/10824669.2020.1757454

Gao, X., Li, P., Shen, J., & Sun, H. (2020). Reviewing assessment of student learning in interdisciplinary STEM education. *International Journal of STEM Education, 7*(1). https://doi.org/10.1186/s40594-020-00225-4

Gardner, J., Harlen, W., Hayward, L., Stobart, G., & Montgomery, M. (2010). *Developing teacher assessment*. McGraw-Hill Education (UK).

Goe, L., Bell, C., & Little, O. (2008). *Approaches to evaluating teacher effectiveness: A research synthesis NCCTQ*. National Comprehensive Center For Teacher Quality.

Harlen, W. (2007). *Assessment of learning*. Sage.

Hattie, J. (2009). *Visible learning: A synthesis of over 800 meta-analyses relating to achievement*. Taylor & Francis.

Herppich, S., Praetorius, A. -K., Förster, N., Glogger-Frey, I., Karst, K., Leutner, D., Behrmann, L., Böhmer, M., Ufer, S., Klug, J., Hetmanek, A., Ohle, A., Böhmer, I., Karing, C., Kaiser, J., & Südkamp, A. (2018). Teachers' assessment competence: Integrating knowledge-, process-, and product-oriented approaches into a competence-oriented conceptual model. *Teaching and Teacher Education, 76*, 181–193. https://doi.org/10.1016/j.tate.2017.12.001

Klein, P. (2012). *Predictive relationships of teacher efficacy, geometry knowledge for teaching, and the cognitive levels of teacher practice on student achievement*. University of Louisville. https://doi.org/10.18297/etd/763

Klieme, E., Pauli, C., & Reusser, K. (2009). The Pythagoras study: Investigating effects of teaching and learning in Swiss and German mathematics classrooms. In T. Janik & T. Seidel (Eds.), *The power of video studies in investigating teaching and learning in the classroom* (pp. 137–160). Waxmann.

Ko, J., & Sammons, P. (2013). *Effective teaching: A review of research and evidence*. CfBT Education Trust.

Korpershoek, H., Harms, T., de Boer, H., van Kuijk, M., & Doolaard, S. (2016). A meta-analysis of the effects of classroom management strategies and classroom management programs on students' academic, behavioral, emotional, and motivational outcomes. *Review of Educational Research, 86*(3), 643–680. https://doi.org/10.3102/0034654315626799

Lipowsky, F., Rakoczy, K., Pauli, C., Drollinger-Vetter, B., Klieme, E., & Reusser, K. (2009). Quality of geometry instruction and its short-term impact on students' understanding of the pythagorean theorem. *Learning and Instruction, 19*(6), 527–537. https://doi.org/10.1016/j.learninstruc.2008.11.001

Marder, J., Thiel, F., & Göllner, R. (2023). Classroom management and students' mathematics achievement: The role of students' disruptive behavior and teacher classroom management. *Learning and Instruction, 86*, 101746. https://doi.org/10.1016/j.learninstruc.2023.101746

Muijs, D., & Reynolds, D. (2017). *Effective teaching: Evidence and practice*. Sage.

Mullis, I. V. S., & Martin, M. O. (Eds.). (2017). *TIMSS 2019 Context questionnaire framework*. TIMSS & PIRLS International Study Center. http://timssandpirls.bc.edu/timss2019/frameworks/framework-chapters/science-framework/

Panadero, E., Jonsson, A., & Botella, J. (2017). Effects of self-assessment on self-regulated learning and self-efficacy: Four meta-analyses. *Educational Research Review, 22*, 74–98. https://doi.org/10.1016/j.edurev.2017.08.004

Perry, L.B., Thier, M., Beach, P. (2023). Opportunities and conditions to learn (OCL): A conceptual framework. *Prospects*. https://doi.org/10.1007/s11125-023-09637-w

Popham, W. J. (1999). *Classroom assessment: What teachers need to know*. Allyn and Bacon.

Porter, A. C. (2002). Measuring the content of instruction: Uses in research and practice. *Educational Researcher, 31*(7), 3–14. https://doi.org/10.3102/0013189X031007003

Praetorius, A.-K., Klieme, E., Herbert, B., & Pinger, P. (2018). Generic dimensions of teaching quality: The German framework of three basic dimensions. *ZDM Mathematics Education, 50*(3), 407–426. https://doi.org/10.1007/s11858-018-0918-4

Schildkamp, K., van der Kleij, F. M., Heitink, M. C., Kippers, W. B., & Veldkamp, B. P. (2020). Formative assessment: A systematic review of critical teacher prerequisites for classroom practice. *International Journal of Educational Research, 103*, 101602. https://doi.org/10.1016/j.ijer.2020.101602

Schlesinger, L., Jentsch, A., Kaiser, G., König, J., & Blömeke, S. (2018). Subject-specific characteristics of instructional quality in mathematics education. *ZDM Mathematics Education, 50*, 475–490. https://doi.org/10.1007/s11858-018-0917-5

Schmidt, W. H., McKnight, C. C., Valverde, G., Houang, R. T., & Wiley, D. E. (Eds.). (1997). *Many visions, many aims: A cross-national investigation of curricular intentions in school mathematics* (Vol. 1). Springer Science & Business Media.

Schmidt, W. H., Guo, S. W., & Houang, R. T. (2021). The role of opportunity to learn in ethnic inequality in mathematics. *Journal of Curriculum Studies, 53*(5), 579–600. https://doi.org/10.1080/00220272.2020.1863475

Senden, B., Nilsen, T., & Blömeke, S. (2022). Instructional quality: A review of conceptualizations, measurement approaches, and research findings. *Ways of Analyzing Teaching Quality: Potentials and Pitfalls*, 140–172.

Senden, B., Nilsen, T., & Teig, N. (2023). The validity of student ratings of teaching quality: Factorial structure, comparability, and the relation to achievement. *Studies in Educational Evaluation, 78*. https://doi.org/10.1016/j.stueduc.2023.101274.

Sigurjónsson, J. Ö. (2023). Quality in Icelandic mathematics teaching: Cognitive activation in mathematics lessons in a Nordic context. [Doctoral dissertation, University of Iceland]. Opinvisindi. https://hdl.handle.net/20.500.11815/3843

Teig, N., Scherer, R., & Nilsen, T. (2019). I know I can, but do I have the time? The role of teachers' self-efficacy and perceived time constraints in implementing cognitive-activation strategies in science. *Frontiers in Psychology, 10*(1697). https://doi.org/10.3389/fpsyg.2019.01697

Teig, N., & Nilsen, T. (2022). Profiles of instructional quality in primary and secondary education: Patterns, predictors, and relations to student achievement and motivation in science. *Studies in Educational Evaluation, 74*, 101170. https://doi.org/10.1016/j.stueduc.2022.101170

Teig, N., Scherer, R., & Olsen, R. V. (2022). A systematic review of studies investigating science teaching and learning: over two decades of TIMSS and PISA. *International Journal of Science Education, 44*(12), 2035–2058. https://doi.org/10.1080/09500693.2022.2109075

Wallace, M. R. (2009). Making sense of the links: Professional development, teacher practices, and student achievement. *Teachers College Record, 111*(2), 573–596. https://doi.org/10.1177/016 146810911100205

Nani Teig is an associate professor at the University of Oslo, Norway. Her research focuses on inquiry-based teaching, scientific reasoning, teaching quality, and academic resilience. She integrates multilevel analyses using data from videos, surveys, assessments, and computer log files. Dr. Teig has received several awards and fellowships, including the Global Education Award, Bruce H. Choppin Dissertation Award, Young CAS Fellow, and UNESCO GEM Fellow.

Trude Nilsen is a research professor at the University of Oslo. She is a leader of Strand 1 at CREATE—Centre for Research on Equality in Education, a leader of the research group LEA, and of funded research projects. She has been engaged as an international external expert for IEA's TIMSS and for OECD's TALIS studies. Her research focuses on teaching quality, educational equality, school climate, and applied methodology including causal inferences.

Kajsa Yang Hansen research concerns educational quality and equity from a comparative perspective. She tackles these issues by investigating students' social, motivational, and cognitive factors in the contexts of their schools, changing societies and education systems. Dr. Yang Hansen also has an interest in analytical techniques for large-scale survey data, e.g., multi-level analysis, Structural Equation Modelling (SEM) and second-generation SEM.

Chapter 3
Analytical Framework

Trude Nilsen◉ **and Nani Teig**◉

3.1 Introduction

The present book adopts a common approach in terms of theories, conceptualizations, and methodology throughout all chapters. The primary objective of this chapter is to describe the *common* methodology that serves as a foundation for each chapter. While the individual chapters can be read independently, it is important to understand that they all share common methodologies and assumptions, which will be thoroughly outlined in this chapter. For instance, a key aspect of the methodology is the uniformity in data preparation across all chapters. Furthermore, this chapter also aims to provide a comprehensive overview of the Trends in International Mathematics and Science Study (TIMSS) and its design. As we progress through the chapter, the overall methodology employed in the book will be examined in greater detail. This examination will cover various aspects of the methodology, including the reliability and validity of the constructs examined in the chapters, the process of data preparation, and the analytical approaches employed. By presenting a clear and coherent understanding of the shared methodology, readers will be better equipped to appreciate the interconnectedness of the chapters and the holistic approach taken in this book.

T. Nilsen (✉)
Department of Teacher Education and School Research, CREATE—Centre for Research on Equality in Education (project number 331640), University of Oslo, Postboks 1099, Blindern, 0317 Oslo, Norway
e-mail: trude.nilsen@ils.uio.no

N. Teig
Department of Teacher Education and School Research, University of Oslo, Postboks 1099 Blindern, 0317 Oslo, Norway
e-mail: nani.teig@ils.uio.no

© The Author(s) 2024   35
N. Teig et al. (eds.), *Effective and Equitable Teacher Practice in Mathematics and Science Education*, IEA Research for Education 14,
https://doi.org/10.1007/978-3-031-49580-9_3

3.2   About TIMSS

3.2.1   The TIMSS Assessment and Questionnaires

TIMSS was first implemented in 1995, following the existence of earlier studies like the First and Second International Mathematics Study (FIMSS and SIMS) (Brown, 1996). TIMSS is made possible through the International Association for the Evaluation of Educational Achievement (IEA), with its administration managed by the TIMSS & PIRLS International Study Centre, Boston College. TIMSS measures students' competence in mathematics and science and is conducted every four years, primarily targeting fourth and eighth-grade students. However, since 2015, Norway shifted the target grades from grade four to grade five and from grades eight to nine to enable better comparisons with other Nordic countries. This was necessary because Norwegian students were typically younger than students in these countries by approximately one year, and adjusting the target grades would help to ensure a fairer comparison (Bergem et al., 2016).

The mathematics test contains more than 200 tasks (hereafter referred to as items), while the science test includes about 250 items (Mullis & Martin, 2017a). Approximately half of these items are in multiple-choice, and the rest are in open-ended format. The test frameworks are based on the participating countries' curricula. All the countries provide feedback to the frameworks and participate in the development of the mathematics and science items in every cycle. The frameworks, test items, and questionnaires are subject to an extensive quality assurance process, which includes several rounds of feedback and revisions as well as a field-trial study (Cotter et al., 2020).

The TIMSS 2019 mathematics assessment for fourth-grade students is divided into three content domains: number, measurement and geometry, and data (Mullis & Martin, 2017a). These domains represent 50%, 30%, and 20% of the test, respectively. Each domain has specific objectives for students to achieve. For instance, in the number domain, students should be able to add and subtract up to 4-digit numbers in simple contextual problems. Similarly, the science assessment for fourth grade also has three content domains: life science (45%), physical science (35%), and earth science (20%), each with its own set of specifications. For example, life science covers a topic about human health, which requires students to be able to identify or describe some methods of preventing disease transmission (e.g., vaccination, washing hands, avoiding people who are sick) and recognize common signs of illness (e.g., high body temperature, coughing, stomach-ache). The framework for the test also specifies a detailed description of the content and objectives for each domain.

In addition to the content domains, the frameworks for mathematics and science assessments also have a cognitive domain that specifies the thinking processes to be assessed (Martin et al., 2017). This domain is divided into three areas: knowing (40%), applying (40%), and reasoning (20%) in the fourth-grade assessment. This cognitive dimension ensures that all aspects of competence are assessed, as students are expected to demonstrate their knowledge of the topic, apply this knowledge to different contexts, and engage in reasoning through processes such as synthesis, evaluation, and generalization.

In addition to the mathematics and science assessments, TIMSS also gathers information related to teaching and learning processes through questionnaires (Mullis & Martin, 2017b). The questionnaires are administered to students, teachers, and school leaders (principals) in grades four and eight, as well as to parents in grade four. The student questionnaires include, among other things, questions about socioeconomic status (SES), minority status (in terms of language), bullying, perceptions of teaching quality, and motivation to learn mathematics and science. Teachers are asked a number of questions, including their educational background, teaching experience and specialization, what they teach in the classroom (content coverage), how they teach and assess their students, and perceptions of the school environment. Similarly, principals are asked, among other things, about their educational background, school composition, instructional resources, and learning environment. Parents answer questions related to their children's education, including educational resources at home and their child's early numeracy and literacy. These questionnaires provide valuable contextual information that can help to better understand the factors that influence student achievement in mathematics and science (Mullis & Martin, 2017b).

3.2.2   The TIMSS Design

International large-scale assessments (ILSAs), such as TIMSS, have complex designs that require special consideration that should be taken into account when analyzing the data.

Hierarchical Design

The target population in TIMSS typically includes students in the fourth and eighth grades, and representative samples are drawn from these populations in each participating country. To achieve this, TIMSS employs a hierarchical design in its sampling and data collection process (Martin et al., 2020). The sampling procedure consists of a two-stage random sample design, which involves selecting schools and then selecting one or more classrooms within these schools (Martin et al., 2020). In addition to the selected students in an intact classroom, the sample also includes their mathematics and science teachers, their principals, and their parents (parents are included only in grade four). While the teachers of the students do not constitute a representative sample, each student is linked to their mathematics and science teachers, which contributes to the validity of inferences related to teachers. This hierarchical design ensures that the data collected is representative of the target population and provides valuable insights into the mathematics and science achievement of students in participating countries.

The hierarchical design has implications for data analysis. If this design is not accounted for in the analyses, the standard errors of the estimates may be underestimated (Rutkowski et al., 2010). This is because students within the same classroom or school tend to resemble one another, hence, violating the requirement of a random sample (Rutkowski et al., 2010). To address this issue, several methods can be employed. The empirical chapters in this book apply multi-level analysis to account for the clustering of the data.

Sampling Weight

Sampling weights are used to adjust the data so that it accurately represents the population being studied and ensures that the data collected from the sample is representative of the entire population (Meinck & Vandenplas, 2020). Without weighting, the data may be biased and may not provide accurate estimates of population parameters. In TIMSS, appropriate sampling weights are necessary to account for the complex sample design and to provide accurate estimates of population parameters (Martin et al., 2020). Researchers need to use the appropriate sampling weights in their analyses to account for the hierarchical design of the sample and to ensure that the results are representative of the population being studied (Rutkowski et al., 2010; Stapleton, 2013). Failure to use appropriate sampling weights can lead to biased estimates of population parameters, which can compromise the validity of the study findings (Meinck & Vandenplas, 2020).

In this book, TIMSS sampling weights and weight factors were taken into account when analyzing the data at the individual or classroom level in the empirical chapters. For analyses using a multilevel model, the chapters use multilevel weights following the recommendations from Rutkowski et al. (2010) and Stapleton (2013). At the student level, the weight is set to a product of student response adjustment and student weight factor (WGTADJ3 × WGTFAC3). At the classroom level, the weight is a product of school response adjustment, school weight factor, classroom response adjustment, and classroom weight factor (WGTADJ1 × WGTFAC1 × WGTADJ2 × WGTFAC2). For more information on weights and weight factors, see LaRoche et al. (2020).

Trend Design

TIMSS is designed to allow for comparisons of student performance in grades four and eight across different cycles of the assessment, which is known as the trend design. This approach allows for the tracking of changes and trends every four years. This is made possible by retaining approximately half of the test items from one cycle to the next, ensuring continuity in the content assessed (Martin & Mullis, 2019). Furthermore, most countries participate in every cycle of TIMSS, which enables the concurrent calibration of scales (Martin & Mullis, 2019). This method allows researchers to scale achievements on the same scale as previous cycles using trend items and trend countries, making direct comparisons across cycles possible (Martin & Mullis, 2019). The trend design helps in identifying changes in student achievement over time, as well as examining the impact of factors, such as content coverage and teaching practices on student performance. However, few take advantage of the trend design in secondary analyses. In this book, the trend design is utilized to examine how changes in content coverage, teaching quality, and assessment practice are related to the changes in achievements over time (see Chaps. 6 and 7).

Plausible Values

As in other ILSAs, TIMSS uses plausible values to represent student proficiency in mathematics and science. In each subject, there are over 200 test items that are used to thoroughly assess mathematics and science (Martin et al., 2017). To minimize the time students spend on the test, items are divided into blocks, preserving the same distribution across content and cognitive domains as the overall test, following the assessment framework (Martin et al., 2020). In TIMSS 2019, there were 28 blocks of items, which comprised 16 blocks of trend items from previous cycles (eight blocks in mathematics and eight blocks in science) and 12 blocks of items that were new in 2019. TIMSS 2015 and 2011 had the same design with 28 blocks. Each student receives two blocks of mathematics items and two blocks of science items (Martin et al., 2020). As individual students only receive a subset of the entire test, plausible values estimate group content-related scale scores for the population, rather than providing accurate individual-level scores (von Davier et al., 2009).

From TIMSS 2011 to 2019, five plausible values were drawn for each student. These values are randomly drawn from an empirically derived distribution of score values based on the student's observed responses to assessment items and selected background variables (von Davier et al., 2009). When analyzing data, researchers must consider these plausible values to ensure accurate estimates of the relationships between variables (Rutkowski et al., 2010; von Davier et al., 2009). This procedure requires separate analyses for each set of plausible values. Once analyses are conducted for all sets, the resultant model parameters are pooled (Laukaityte & Wiberg, 2017). The pooling involves calculating the means across all sets of model parameters, while the variances are quantified according to Rubin's combination rules, which consider the variances within and between plausible values and the number of plausible values (Laukaityte & Wiberg, 2017). The empirical chapters in this book employed Mplus, a statistical analysis software, which offers a convenience option (i.e., TYPE = IMPUTATION) to perform analysis for each set of plausible values and automatically combines the resultant model parameters.

Cross-sectional Design

TIMSS, along with other ILSAs, employs a cross-sectional design (Martin et al., 2017). This design involves collecting data at a specific point in time, typically once every four years in the case of TIMSS, to assess and compare the performance of students across participating countries. In a cross-sectional design, data are collected from different participants in the same age group or grade level, without following them over time. This approach allows researchers to identify patterns, trends, and relationships between various factors, such as education systems, teaching practices, and student achievement. However, since data are collected only at a single time point, it is not possible to establish causal relationships or determine why some variables change over time (Cummings, 2018). As a result, causal language should be avoided in favor of discussing "relationships" rather than "effects". For instance, instead of discussing the "effect" of content coverage on achievement, it is more appropriate to refer to it as the "relationship" between content coverage and achievement.

Given that this book is intended for educational policy stakeholders, practitioners, and researchers, using overly technical language may hinder clarity and comprehension, particularly when discussing advanced methodologies. As a result, some language choices in this book may be simplified to improve understanding. Nevertheless, it is crucial to emphasize that causal relationships between predictors and outcomes cannot be established through cross-sectional designs.

3.3   The Main Measures Used in This Book

This section focuses on describing common measures of teacher practice examined throughout the book (for the theoretical foundations of teacher practice, see Chap. 2).

3.3.1   What Teachers Teach: Content Coverage

Content coverage represents a critical aspect of any curriculum, as it outlines the topics students will learn and the knowledge they are expected to acquire. TIMSS distinguishes between the intended curriculum at the national level, the implemented curriculum at the classroom level, and the attained curriculum as learning outcomes at the student level (Mullis & Martin, 2017b). This book mostly focuses on a narrow concept of content coverage to describe student exposure to various topics in mathematics and science. Specifically, content coverage refers to the *TIMSS' mathematics and science topics* implemented by the teachers in classrooms using the TIMSS teacher questionnaire. This measure of content coverage is employed in Chaps. 4, 6, and 8. In addition, Chap. 4 examines the alignment between content coverage (the implemented curriculum) with the topics covered by the intended national curriculum in the Nordic countries (the curriculum questionnaire) and the attained curriculum as assessed in the TIMSS test. Meanwhile, Chaps. 6 and 8 investigate the relations between content coverage and student achievement across the various domains of mathematics and science (the attained curriculum).

Content coverage was assessed using the teacher questionnaire, focusing on the extent to which teachers had taught specific topics covered in the TIMSS test to fourth-grade students (Martin et al., 2020). Teachers were asked to rate when they taught various topics within the subdomains of mathematics and science. In TIMSS 2019 for mathematics, there were seven topics in the content domain number, seven in measurement and geometry, and three in data, whereas, in science, there were seven topics in life science, 12 in physical science, and seven in earth science. For example, in mathematics, for the number domain, teachers were asked when the class had been taught the topic "concepts of whole numbers, including place value and ordering". The response scale includes mostly taught before this year, mostly taught this year, and not yet taught or just introduced. There are two main issues with this scale. First, new teachers may not know what students have been taught before. Second, the relatively large number of items may lead to a higher rate of missing data, particularly in science (which is asked after mathematics in the questionnaire). Furthermore, "not yet taught" and "just introduced" represent distinct concepts.

Given the challenges with the response scale, it was challenging to measure content coverage as an indicator of the implemented curriculum. To address this issue, this book used the percentages of students who had been taught each of the topics (before or during the school year) as reported by their teachers, averaged across all topics in each subject domain, and also across all topics in all subject domains. In mathematics, these percentages are represented by three variables: the percentages of students taught the topics number (ATDMNUM), measurement and geometry (ATDMGEO), and data (ATDMDAT). In science, the percentages of students taught the topics life science (ATDSLIF), physical science (ATDSPHY), and earth science (ATDSEAR) were used. It is worth noting that the missing values at the item level were quite large, especially in science. In TIMSS 2019, data for the percentages of students taught science topics are available for at least 70% but less than 85% of the students in Denmark and Sweden, whereas, in Norway, the data are available for at least 50 percent but less than 70% of the students. The high proportion of missing values in these countries highlights the need for caution when interpreting the results, as they may not fully reflect the true content coverage.

In the curriculum questionnaire, TIMSS' National Research Coordinators (NRC) responded to a set of questions focusing on national curriculum policies and practices related to each country's education system, as well as the organization and content of mathematics and science curricula (Martin et al., 2020). The NRCs were asked whether each of the TIMSS mathematics and science topics were included in their countries' intended curriculum and, if so, whether the topics were intended to be taught to "all or almost all students" or "only the more able students". The TIMSS 2019 curriculum questionnaire was administered online, and participants were advised to draw on the expertise of curriculum specialists and educators. However, there were no reliability checks or procedures in place to reduce measurement uncertainty or improve reliability (Martin et al., 2020). Consequently, the curriculum questionnaire was examined together with the teacher questionnaire in this book.

To measure the attained curriculum, student achievement in the specific content domains was used. For mathematics, five plausible values of achievement were used for number, measurement and geometry, and data, similarly five plausible values were used for life science, physical science, and earth science for science.

3.3.2 How Teachers Teach: Teaching Quality

TIMSS measures teaching quality using a combination of student and teacher questionnaires. As discussed in Chap. 2 this book used three basic dimensions of teaching quality consisting of classroom management, supportive climate, and cognitive activation (Klieme et al., 2009). *Classroom management* was assessed for the first time in TIMSS 2019 using student questionnaires that asked students how often various situations happened in their mathematics classrooms. Students were presented with six items (e.g., "my teacher has to keep telling us to follow the classroom rules" or "students interrupt the teacher") and were asked to indicate the frequency of their occurrence using a response scale that includes: never, some lessons, about half the lessons, and every or almost every lesson. A scale called Disorderly Behavior During Mathematics Lessons was also created based on students' responses to these items.

Supportive climate encompasses various aspects, including teacher support, classroom interaction (teacher-student and student–student relationships), and instructional clarity. This book focuses on the *teacher support and instructional clarity* that was assessed using the student questionnaires in TIMSS 2011 to 2019. It measured student agreement on various statements with a response scale that includes agree a lot, agree a little, disagree a little, and disagree a lot. Note that only two items were similar across these cycle (i.e., "I know what my teacher expects me to do" and "my teacher is easy to understand"). The items were situated separately in mathematics and science classrooms.

This book differentiates between general and subject-specific *cognitive activation* (see Chap. 2 for further details). Teachers reported how often they engaged students in various activities with a response scale that ranges from never to every or almost every lesson. Few items are similar across TIMSS 2011 to 2019; two items in generic cognitive activation (e.g., "relate the lesson to students' daily lives"), three items in mathematics cognitive activation (e.g., "apply what students have learned to new problem situations on their own"), and five items in science cognitive activation (e.g., "design or plan experiments or investigations"). For the first time in 2019, TIMSS added an item to the student questionnaire and asked students how often they conduct experiments in their science lessons with a response scale of never, a few times a year, once or twice a month, and at least once a week. This item represents inquiry-based cognitive activation in science and is included in the analysis to supplement the teacher questionnaire.

The TIMSS' hierarchical design, which involves clustering students in intact classrooms and collecting information from both student and teacher questionnaires, is highly suitable for measuring teaching quality. A significant advantage of this method is that students' responses provide first-hand experiences of the teaching process. Ideally, if the goal is to accurately measure teaching quality, all students would rate their teacher similarly. However, perceptions may vary among students. To account for these variations, the chapters that explore whether teaching quality may "explain" achievement differences between classrooms employ a two-level model at the student and classroom levels. In this model, the perceptions of teaching quality are controlled at the student level, and the results are focused on the classroom level. This approach aligns with previous research and offers a reliable method for measuring teaching quality (e.g., Lüdtke et al., 2007; Marsh et al., 2009).

Some challenges may arise in measuring teaching quality using student and teacher questionnaires. Research has shown that young students may have difficulties in evaluating their teachers and distinguishing between different aspects of teaching quality (Lüdtke et al., 2007). This means that students may perceive teachers who generally perform well as having high quality in all aspects of teaching, regardless of whether or not this is the case. Conversely, teachers who perform poorly in one aspect may be perceived as having low quality in all aspects of teaching (Teig & Nilsen, 2022). TIMSS also collects information about teaching quality through the teacher questionnaire that covers more items and deeper aspects of teaching quality than the student questionnaire. Nevertheless, these self-reported measures may be susceptible to social desirability bias (Muijs, 2006). Teachers may feel pressure to provide responses that they believe are socially desirable, rather than providing honest answers. This can lead to the over-reporting of positive behaviors and under-reporting of negative behaviors. This book used both the student and teacher questionnaires to minimize the challenges associated with both approaches.

As previously discussed, the TIMSS's design, which links students with their teachers, is well-suited for measuring teacher practice, and indeed, teacher practices have been measured in all TIMSS cycles since 1995 (Klieme & Nilsen, 2022). However, the measurement of teaching quality—tailored specifically to classroom management, teacher support and instructional clarity, and cognitive activation—is a more recent inclusion. These aspects of teaching quality were not specifically incorporated into the TIMSS context questionnaire framework until 2015. The aspect of classroom management was later added in 2019 but only for mathematics. Moreover, with each TIMSS cycle, more information is gathered, leading to valuable insights and improvements in the measures for teaching quality. Therefore, the quality of the teaching quality measures in TIMSS has improved through both pilot studies and the main studies in 2015, 2019, and 2023 (Klieme & Nilsen, 2022). This implies that the validity of teaching quality is higher in the chapters that utilize data from TIMSS 2019 (i.e., Chaps. 5 and 9) than in the trend chapter that analyses changes from 2011 to 2019 (i.e., Chap. 7).

3.3.3 How Teachers Assess Their Students: Assessment Practice

TIMSS has measured various aspects of teachers' assessment practices in 2011, 2015, and 2019. Nevertheless, only items related to homework were similar across these cycles. Measuring homework in TIMSS is challenging due to varying definitions and frequencies across countries and schools. Two primary homework measures exist: the frequency of assigned homework and the expected time students spend on it. Interpreting these measures separately can be difficult because teachers may allocate homework rarely but provide tasks that would take the students a long time to complete. However, combining the frequency and time spent on homework provides a more useful estimate (see Chap. 7 for a description of the procedure). In addition, teachers were also asked how they integrate homework into their teaching (in-class homework discussion). Teachers reported on how often they correct assignments, provide feedback, discuss homework in class, and monitor the completion of the homework. This additional information helps create a more comprehensive understanding of homework as an assessment practice in different educational contexts.

In TIMSS 2019, five new items were added to measure how much importance teachers place on various assessment strategies in mathematics and science, such as observing students, asking questions during class, short written assessments, longer tests, and long-term projects. These new items allow for a more in-depth analysis of teachers' assessment practices and contribute to a better understanding of their impact on students' learning outcomes.

3.4 Analytical Approaches

3.4.1 Data Preparation

To maintain coherence and enable comparisons across the chapters of this book, all data were prepared in advance and analyzed in the same manner. Two main files were created: one for the chapters that analyze TIMSS 2019 data, and another file containing merged data from 2011, 2015, and 2019 for the chapters conducting trend analyses. The IDB Analyzer was employed to merge teacher and student data and to combine data from different countries. Some variables required reverse coding to ensure that higher values represented more positive outcomes.

To accommodate multi-level analyses at the student and classroom levels, data that include multiple teachers per student (between 2 and 6%) were deleted randomly so that each student was only linked to one mathematics and one science teacher. This simplification helps to provide a clear, one-to-one relationship between each student and their respective teachers. This straightforward link improves the clarity and interpretability of the results, as it avoids potential bias from averaging responses across different teachers, who may have different teaching practices and interactions with the student. It also eliminates the potential issue of students having varying experiences with different teachers, which could complicate the interpretation of the findings. For example, if one teacher interacts with the student more than another, it may not be appropriate to weigh their responses equally. This approach also adheres to the nested structure assumed in multi-level models, where students are nested within teachers. This approach is taken to maintain clear hierarchical data analyses, even though it necessitates the removal of some data (between 2 and 6%).

For the trend analysis that required merged data from the three TIMSS cycles, a dummy variable called TIME was created and coded as 0 for 2011, 1 for 2015, and 2 for 2019. This variable allowed for an easy comparison of trends across different cycles and aided in the identification of patterns or changes over time. Furthermore, unique identification numbers were assigned to students and teachers to guarantee uniqueness across countries and over time. This approach facilitated the tracking of individual data points and the comparison of trends across different cycles. By ensuring the uniqueness of identification numbers, potential issues related to data duplication or misinterpretation were minimized, contributing to the overall reliability and validity of the findings.

3.4.2   Preliminary Analysis

In all empirical chapters, we employed Mplus software to analyze the data within the Structural Equation Modeling (SEM) framework. SEM consists of two parts: a measurement part, which assesses the reliability and validity of constructs and the model itself, and a structural part, which consists of regression between one or more variables or constructs. The measurement part of the modelling includes Confirmatory Factor Analysis (CFA). Whenever possible or appropriate, CFA was used to create latent variables, which represent the underlying constructs that are not directly observable (Brown, 2015). CFA is a valuable approach to assessing the reliability and validity of a latent variable. Using Mplus, we can estimate the so-called factor loadings (with values between 0 and 1) to ascertain how well each item in a latent variable measures the underlying concept. For instance, if we want to measure student motivation in mathematics, items like: "I like mathematics" or "mathematics is my favorite subject" would typically present high factor loadings (e.g., 0.9). Conversely, if we included an item on whether the students like chocolate, the factor loading for this item would be very low and probably insignificant. Thus, careful item selection based on relevant theories in CFA is crucial to create an accurate representation of the constructs under investigation.

SEM integrates factor analysis, path analysis, and regression techniques, allowing researchers to test and estimate the relationships between observed and latent variables (Brown, 2015). SEM is a robust methodology that enables researchers to perform estimations simultaneously at both the student and classroom levels (Hox et al., 2017; Morin et al., 2014). This approach allows for a more comprehensive analysis of the complex relationships among variables in educational research. Keys in this process are model fit and factor loadings, which serve as indicators of reliability and validity, supporting the overall quality of the findings. By assessing model fit, researchers can determine how well the proposed model represents the observed data (Brown, 2015; Hu & Bentler, 1999). Factor loadings, on the other hand, show the strength of the relationships between observed variables and their underlying latent constructs (Brown, 2015). These assessments are crucial for establishing the credibility of the results and for interpreting the implications of the findings in the context of educational research.

Before conducting the main analyses within the chapters, it was crucial to determine if the measures were comparable across countries. Therefore, Measurement Invariance (MI) testing was conducted to determine whether the constructs of teacher practices measure the same underlying construct across different groups. MI testing is essential in cross-cultural research to ensure that comparisons made between groups are meaningful and valid. Note that the MI testing was only conducted to ensure comparison across Nordic countries for the TIMSS 2019 data. The MI testing across time was not conducted due to the very few similar items that measure teacher practice across TIMSS 2011 to 2019.

The MI testing was conducted within the framework of SEM or CFA. The process involves a series of hierarchical model comparisons, where increasingly restrictive models are compared to assess the invariance of the measurement. There are three primary levels of measurement invariance (Kang et al., 2015; Sass & Schmitt, 2013):

(1) Configural invariance. This level establishes that the same factor structure (i.e., the relationship between items and the underlying latent construct) holds across groups. In this stage, no equality constraints are imposed on the model parameters.
(2) Metric invariance (weak invariance). At this level, the factor loadings (the strength of the relationship between items and the latent construct) are constrained to be equal across groups. Establishing metric invariance suggests that the units of measurement are the same across groups, allowing for meaningful comparisons of relationships between constructs.
(3) Scalar invariance (strong invariance). This level tests whether the item intercepts (the point at which the item is expected to have a zero score) are equal across groups. Scalar invariance is necessary for meaningful comparisons of latent means or group differences.

To assess the fit of the invariance models, fit indices such as the Comparative Fit Index (CFI), Tucker-Lewis Index (TLI), Root Mean Square Error of Approximation (RMSEA), and the Standardized Root Mean Square Residual (SRMR) were used. If the fit of a more restrictive model was not substantially worse than the fit of the previous model, it could be concluded that the measurement is invariant at that level (Chen, 2007; Cheung & Rensvold, 2002). When comparing the nested models, the following cut-off values for fit indices were used to determine the degree of invariance achieved: a decrease of 0.01 or less in the CFI or TLI (Cheung & Rensvold, 2002), a change in RMSEA of 0.015 or less, and an increase in SRMR of 0.03 or less (Chen, 2007).

As shown in Appendix 1, all constructs of teacher practice exhibited metric or scalar invariance, with the exception of cognitive activation strategies in mathematics using the student questionnaire, in-class homework discussion, and teachers' emphasis on various assessment strategies. Consequently, analyses involving these aspects of teacher practice are considered as manifest variables rather than latent constructs. Moreover, cross-country comparisons that incorporate these variables need to be carefully interpreted. It is essential to acknowledge that these non-invariant constructs might reflect differences in understanding or interpretation across countries. Caution should be exercised when drawing conclusions based on these variables, as the observed differences may not necessarily indicate true differences in the constructs themselves.

By using a consistent analytical approach across chapters, this book offers a comprehensive and coherent view of teacher practice. The utilization of MI testing, SEM, and CFA analyses allows for reliable and valid comparisons across countries, contributing to the overall understanding of various aspects of teacher practice in different educational contexts. This approach strengthens the book's capacity to provide insights and inform policy discussions, helping stakeholders make informed decisions for improving educational practices and outcomes.

3.5 Concluding Remarks

To facilitate the comparison of findings across the chapters, it was crucial for all authors to use the same prepared data and adopt a consistent methodology. However, certain deviations in operationalizing the constructs were occasionally needed. For instance, authors might have made minor adjustments to the operationalizations to enable model convergence or enhance model fit. Furthermore, due to changes in some items across the 2011 to 2019 cycles, the constructs used in chapters analyzing trends might differ from those in chapters utilizing TIMSS 2019 data.

Regardless of these minor discrepancies, the overall coherent approach across the chapters supports a valuable integration of knowledge about what teachers teach (content coverage), how teachers teach (teaching quality), and how teachers assess their students (assessment practice) from a range of perspectives. This uniformity ensures that the findings from different chapters can be effectively compared, offering a comprehensive understanding of the intricate relationship between teacher practice and student achievement in various educational contexts.

Appendices

Appendix 1 The Results of Measurement Invariance Across Nordic Countries Based on TIMSS 2019

Teaching quality: Classroom management (student questionnaire)—Mathematics

Items: How often do these things happen in your mathematics lessons?

1. Students don't listen to what the teacher says
2. There is disruptive noise
3. It is too disorderly for students to work well
4. My teacher has to wait a long time for students to quiet down
5. Students interrupt the teacher
6. My teacher has to keep telling us to follow the classroom rules.

	χ2	df	χ2 diff	df diff	p-value	RMSEA	SRMR	CFI	TLI	Δ RMSEA	Δ SRMR	Δ CFI	Δ TLI
Configural: equal form	940.146	36	–	–	–	0.083	0.023	0.977	0.961	–	–	–	–
Metric: equal factor loadings	1027.717	51	87.571	15	< 0.001	0.072	0.031	0.975	0.97	– 0.011	0.008	– 0.002	0.009
Scalar: equal intercepts	2569.5	66	1541.8	15	< 0.001	0.101	0.055	0.936	0.942	0.018	0.032	– 0.041	– 0.019

Teaching quality: Teacher support and instructional clarity (student question-naire)—Mathematics

Items: How much do you agree with these statements about your mathematics lessons?

1. I know what my teacher expects me to do
2. My teacher is easy to understand
3. My teacher has clear answers to my questions
4. My teacher is good at explaining mathematics
5. My teacher does a variety of things to help us learn
6. My teacher explains a topic again when we don't understand.

	χ2	df	χ2 diff	df diff	p-value	RMSEA	SRMR	CFI	TLI	Δ RMSEA	Δ SRMR	Δ CFI	Δ TLI
Configural: equal form	927.295	36	–	–	–	0.082	0.027	0.969	0.948	–	–	–	–
Metric: equal factor loadings	1012.158	51	84.863	15	< 0.001	0.071	0.032	0.967	0.961	– 0.011	0.005	– 0.002	0.013
Scalar: equal intercepts	1485.35	66	473.19	15	< 0.001	0.042	0.076	0.951	0.955	– 0.040	0.049	– 0.018	0.007

Teaching quality: Teacher support and instructional clarity (student question-naire)—Science

Items: How much do you agree with these statements about your science lessons?

1. I know what my teacher expects me to do
2. My teacher is easy to understand
3. My teacher has clear answers to my questions
4. My teacher is good at explaining mathematics
5. My teacher does a variety of things to help us learn
6. My teacher explains a topic again when we don't understand.

	$\chi 2$	df	$\chi 2$ diff	df diff	p-value	RMSEA	SRMR	CFI	TLI	Δ RMSEA	Δ SRMR	Δ CFI	Δ TLI
Configural: equal form	1054.088	36	–	–	–	0.087	0.026	0.972	0.953	–	–	–	–
Metric: equal factor loadings	1177.004	51	122.92	15	< 0.001	0.077	0.035	0.969	0.964	– 0.010	0.009	– 0.003	0.011
Scalar: equal intercepts	1952.793	66	775.79	15	< 0.001	0.088	0.048	0.948	0.953	0.001	0.022	– 0.024	0.000

References

Bergem, O. K., Kaarstein, H., & Nilsen, T. (2016). TIMSS 2015. In *Vi kan lykkes i realfag [We can suceed in mathematics and science]* (pp. 11–21). Universitetsforlaget. https://doi.org/10.18261/97882150279999-2016-02

Brown, M. (1996). FIMS and SIMS: The first two IEA international mathematics surveys. *Assessment in Education: Principles, Policy & Practice, 3*(2), 193–212. https://doi.org/10.1080/0969594960030206

Brown, T. A. (2015). *Confirmatory factor analysis for applied research*. Guilford publications.

Chen, F. F. (2007). Sensitivity of goodness of fit indexes to lack of measurement invariance. *Structural Equation Modeling: A Multidisciplinary Journal, 14*(3), 464–504. https://doi.org/10.1080/10705510701301834

Cheung, G. W., & Rensvold, R. B. (2002). Evaluating goodness-of-fit indexes for testing measurement invariance. *Structural Equation Modeling, 9*(2), 233–255. https://doi.org/10.1207/S15328007SEM0902_5

Cotter, K. E., Centurino, V. A. S., & Mullis, I. V. S. (2020). Developing the TIMSS 2019 mathematics and science achievement instruments. In M. O. Martin, M. von Davier, & I. V. S. Mullis (Eds.), *Methods and procedures: TIMSS 2019 technical Report* (pp. 1.1–1.36). TIMSS & PIRLS International Study Center, Boston College. https://timssandpirls.bc.edu/timss2019/methods/chapter-1.html

Cummings, C. L. (2018). *Cross-sectional design*. SAGE Publications. https://doi.org/10.4135/9781483381411

Hox, J. J., Moerbeek, M., & van de Schoot, R. (2017). Multilevel analysis: Techniques and applications. *Routledge*. https://doi.org/10.4324/9781315650982

Hu, L. T., & Bentler, P. M. (1999). Cutoff criteria for fit indexes in covariance structure analysis: Conventional criteria versus new alternatives. *Structural Equation Modeling: A Multidisciplinary Journal, 6*(1), 1–55. https://doi.org/10.1080/10705519909540118

Kang, Y., McNeish, D. M., & Hancock, G. R. (2015). The role of measurement quality on practical guidelines for assessing measurement and structural invariance. *Educational and Psychological Measurement, 76*(4). https://doi.org/10.1177/0013164415603764

Klieme, E., Pauli, C., & Reusser, K. (2009). The pythagoras study: Investigating effects of teaching and learning in swiss and german mathematics classrooms. In T. Janik & T. Seidel (Eds.), *The power of video studies in investigating teaching and learning in the classroom* (pp. 137–160). Waxmann.

Klieme, E., & Nilsen, T. (2022). Teaching quality and student outcomes in TIMSS and PISA. In T. Nilsen, A. Stancel-Piątak, & J.-E. Gustafsson (Eds.), (pp. 1089–1134). Springer International Publishing. https://doi.org/10.1007/978-3-030-88178-8_37

LaRoche, S., Joncas, M., & Foy, P. (2020). Sample design in TIMSS 2019. In M. O. Martin, M. von Davier, & I. V. S. Mullis (Eds.), *Methods and procedures: TIMSS 2019 technical report* (pp. 3.1–3.33). TIMSS & PIRLS International Study Center , Boston College. https://timssandpirls.bc.edu/timss2019/methods/chapter-3.html

Laukaityte, I., & Wiberg, M. (2017). Using plausible values in secondary analysis in large-scale assessments. *Communications in Statistics—Theory and Methods, 46*(22), 11341–11357. https://doi.org/10.1080/03610926.2016.1267764

Lüdtke, O., Trautwein, U., Kunter, M., & Baumert, J. (2007). Reliability and agreement of student ratings of the classroom environment: A reanalysis of TIMSS data. *Learning Environments Research, 9*(3), 215–230. https://doi.org/10.1007/s10984-006-9014-8

Marsh, H. W., Lüdtke, O., Robitzsch, A., Trautwein, U., Asparouhov, T., Muthén, B., & Nagengast, B. (2009). Doubly-latent models of school contextual effects: Integrating multilevel and structural equation approaches to control measurement and sampling error. *Multivariate Behavioral Research, 44*(6), 764–802. https://doi.org/10.1080/00273170903333665

Martin, M. O., Mullis, I. V. S., & Foy, P. (2017). TIMSS 2019 assessment design. In I. V. S. Mullis & M. O. Martin (Eds.). TIMSS & PIRLS International Study Center, Boston College. http://tim ssandpirls.bc.edu/timss2019/frameworks/framework-chapters/science-framework/

Martin, M. O., von Davier, M., Mullis, I. V. S., & Foy, P. (2020). *Methods and procedures: TIMSS 2019 technical report.* TIMSS & PIRLS International Study Center, Boston College. https://tim ssandpirls.bc.edu/timss2019/methods

Martin, M. O., & Mullis, I. V. (2019). TIMSS 2015: Illustrating advancements in large-scale international assessments. *Journal of Educational and Behavioral Statistics, 44*(6), 752–781. https://doi.org/10.3102/1076998619882203

Meinck, S., & Vandenplas, C. (2020). Sampling design in ILSA. In T. Nilsen, A. Stancel-Piątak, & J.-E. Gustafsson (Eds.), *International handbook of comparative large-scale studies in education: perspectives, methods and findings* (pp. 1–25). Springer International Publishing. https://doi.org/10.1007/978-3-030-38298-8_25-1

Morin, A. J. S., Marsh, H. W., Nagengast, B., & Scalas, L. F. (2014). Doubly latent multilevel analyses of classroom climate: An illustration. *The Journal of Experimental Education, 82*(2), 143–167. https://doi.org/10.1080/00220973.2013.769412

Muijs, D. (2006). Measuring teacher effectiveness: Some methodological reflections. *Educational Research and Evaluation, 12*(1), 53–74. https://doi.org/10.1080/13803610500392236

Mullis, I. V. S., & Martin, M. O. (Eds.). (2017a). *TIMSS 2019 assessment framework.* TIMSS & PIRLS International Study Center, Boston College. http://timssandpirls.bc.edu/timss2019/fra meworks/framework-chapters/science-framework/

Mullis, I. V. S., & Martin, M. O. (Eds.). (2017b). *TIMSS 2019 context questionnaire Framework.* TIMSS & PIRLS International Study Center, Boston College. http://timssandpirls.bc.edu/tim ss2019/frameworks/framework-chapters/science-framework/

Rutkowski, L., Gonzalez, E., Joncas, M., & von Davier, M. (2010). International large-scale assessment data: Issues in secondary analysis and reporting. *Educational Researcher, 39*(2), 142–151. https://doi.org/10.3102/0013189X10363170

Sass, D. A., & Schmitt, T. A. (2013). Testing measurement and structural invariance. In T. Teo (Ed.), *Handbook of quantitative methods for educational research* (pp. 315–345). SensePublishers. https://doi.org/10.1007/978-94-6209-404-8_15

Stapleton, L. (2013). Incorporating sampling weights into single-and multilevel analyses. In L. Rutkowski, M. von Davier, & D. Rutkowski (Eds.), *Handbook of international large-scale assessment: Background, technical issues, and methods of data analysis* (pp. 363–388). CRC Press. https://doi.org/10.1201/b16061

Teig, N., & Nilsen, T. (2022). Profiles of instructional quality in primary and secondary education: Patterns, predictors, and relations to student achievement and motivation in science. *Studies in Educational Evaluation, 74*, 101170. https://doi-org/10.1016/j.stueduc.2022.101170.

von Davier, M., Gonzalez, E., & Mislevy, R. (2009). What are plausible values and why are they useful. *IERI Monograph Series, 2*(1), 9–36.

Trude Nilsen is a research professor at the University of Oslo. She is a leader of Strand 1 at CREATE—Centre for Research on Equality in Education, a leader of the research group LEA, and of funded research projects. She has been engaged as an international external expert for IEA's TIMSS and for OECD's TALIS studies. Her research focuses on teaching quality, educational equality, school climate, and applied methodology including causal inferences.

Nani Teig is an associate professor at the University of Oslo, Norway. Her research focuses on inquiry-based teaching, scientific reasoning, teaching quality, and academic resilience. She integrates multilevel analyses using data from videos, surveys, assessments, and computer log files. Dr. Teig has received several awards and fellowships, including the Global Education Award, Bruce H. Choppin Dissertation Award, Young CAS Fellow, and UNESCO GEM Fellow.

Chapter 4
Content Coverage: Development Over Time and Correlation with Achievement

Rune Müller Kristensen⊙ and **Victoria Rolfe**⊙

4.1 Introduction

From a didactical perspective, alignment between the intended and implemented curriculum (educational goals at the system and classroom level) and the achieved or attained curriculum (learning outcomes as gained by students) is considered to support students' learning. Such alignment is thus seen as a vital characteristic of effective teacher practice (Daus et al., 2019) and as essential in offering equal opportunities to students across schools. These different levels of content coverage develop continuously, as educational reforms change national curricula, didactical principles develop, and teachers' experiences and the materials they have access to shape how they implement the curriculum over time. For an example of the latter, one need to look no further than how much the use of information technology (IT) in schools has changed between the most recent rounds of TIMSS.[1] However, it can be difficult to compare the three different levels of content coverage as they describe the content in three different ways. A national curriculum describes the intended curriculum in relatively broad and abstract terms, extending across several school years. The implemented curriculum in the form of teachers' descriptions of the content covered when responding to the TIMSS teacher questionnaire's section on

---

[1] The share of Danish fourth grade students with access to their own device in all lessons increased from 32 percent in TIMSS 2015 to 72 percent in TIMSS 2019 (Kjeldsen et al., 2020).

R. M. Kristensen (✉)
Danish School of Education, University of Aarhus, Tuborgvej 164, 2400 Copenhagen, NV, Denmark
e-mail: ruvk@edu.au.dk

V. Rolfe
Department of Education and Special Education, University of Gothenburg, P.O. Box 300, 40530 Gothenburg, Sweden
e-mail: victoria.rolfe@gu.se

N. Teig et al. (eds.), *Effective and Equitable Teacher Practice in Mathematics and Science Education*, IEA Research for Education 14,
https://doi.org/10.1007/978-3-031-49580-9_4

content coverage is based on somewhat technical terms associated with the subject, while their everyday descriptions are presumably closer to the terminology they use during lessons. Finally, the attained curriculum in terms of students' learning outcomes is described empirically, using specific test items within the subject that may not be clearly associated with the different topics within the curriculum.

This chapter presents how these three curriculum levels are measured and the changes that have occurred based on the 2011, 2015, and 2019 cycles of the TIMSS fourth-grade study conducted in the Nordic countries. Furthermore, it comments on the Icelandic curriculum where relevant, as the only Nordic country not participating in TIMSS.

4.2 TIMSS and Curricula in the Nordic Countries

The TIMSS goal of assessing and comparing student learning necessitates close collaboration with representatives from all participating educational systems. In general, the student assessments implemented in TIMSS and described in the TIMSS Assessment Frameworks (Mullis & Martin, 2013, 2017; Mullis et al., 2009) are considered to cover the participating countries' curricula relatively well. However, the overlap between assessment frameworks and the national curriculum varies between countries (Wagner & Hastedt, 2022).

4.2.1 The Test-Curriculum Matching Analysis

To measure how well TIMSS covers a country's curriculum, a Test-Curriculum Matching Analysis (TCMA) was conducted for each country, comparing student performance based on all items included in TIMSS with performance based only on those items that were considered within the country's curriculum. A summary of the TCMA for the four Nordic countries participating in TIMSS is presented in Table 4.1 (mathematics) and Table 4.2 (science) for the years 2011 to 2019. The tables show the number of score points for test items within each domain and the number of score points considered as falling within each country's curriculum in each cycle. In 2011 and 2015, the results of the analyses were presented as the average percentage of correct test-item responses for all test items compared to the average percentage corrected for country specific test items. This was changed in 2019 to the calculated student score based on all test items vs the score based on the test items considered to be within the country's curriculum. Thus, the TCMA analyses cannot be compared before 2019.

Table 4.1 Test-curriculum matching analysis for the years 2011 to 2019 in mathematics

Mathematics	2011	2015	2019
Number of test items in TIMSS	175	169	171
Number of test-item score points in TIMSS	184	178	183
Denmark			
Test items included in national curriculum	170	140	164
Test-item score points in national curriculum	179	146	176
Mathematics achievement based on all items			525 (1.9)
Mathematics achievement on country test items			526 (1.9)
Percentage correct, all test items	59 (0.6)	61 (0.7)	
Percentage correct, national test items	60 (0.6)	56 (0.7)	
Finland			
Test items included in national curriculum	163	156	163
Test-item score points in national curriculum	169	166	175
Mathematics achievement based on all items			532 (2.3)
Mathematics achievement on country test items			532 (2.4)
Percentage correct, all test items	60 (0.6)	55 (0.5)	
Percentage correct, national test items	61 (0.6)	56 (0.5)	
Norway			
Test items included in national curriculum	172	167	157
Test-item score points in national curriculum	181	176	168
Mathematics achievement based on all items			543 (2.2)
Mathematics achievement on country test items			543 (2.2)
Percentage correct, all test items	48 (0.7)	59 (0.7)	
Percentage correct, national test items	49 (0.7)	59 (0.7)	
Sweden			
Test items included in national curriculum	149	130	131
Test-item score points in national curriculum	156	138	142
Mathematics achievement based on all items			521 (2.8)
Mathematics achievement on country test items			525 (2.8)
Percentage correct, all test items	53 (0.5)	51 (0.7)	
Percentage correct, national test items	50 (0.5)	55 (0.7)	

Source Mullis et al. (2012a, 2012b, 2016, 2020). Standard errors in parentheses

Table 4.2 Test-curriculum matching analysis for the years 2011 to 2019 in science

Science	2011	2015	2019
Number of test items in TIMSS	169	168	169
Number of test-item score points in TIMSS	181	180	174
Denmark			
Test items included in national curriculum	164	139	155
Test-item score points in national curriculum	177	149	159
Science achievement based on all items			522 (2.4)
Science achievement on country test items			524 (2.4)
Percentage correct, all test items	54 (0.5)	53 (0.4)	
Percentage correct, national test items	55 (0.5)	58 (0.4)	
Finland			
Test items included in national curriculum	124	107	97
Test-item score points in national curriculum	133	113	99
Science achievement based on all items			555 (2.6)
Science achievement on country test items			555 (2.6)
Percentage correct, all test items	63 (0.4)	58 (0.4)	
Percentage correct, national test items	66 (0.5)	62 (0.4)	
Norway			
Test items included in national curriculum	138	109	143
Test-item score points in national curriculum	149	116	146
Science achievement based on all items			539 (2.2)
Science achievement on country test items			541 (2.2)
Percentage correct, all test items	47 (0.4)	55 (0.5)	
Percentage correct, national test items	47 (0.4)	59 (0.5)	
Sweden			
Test items included in national curriculum	140	101	128
Test-item score points in national curriculum	152	107	131
Science achievement based on all items			537 (3.3)
Science achievement on country test items			539 (3.3)
Percentage correct, all test items	55 (0.5)	56 (0.7)	
Percentage correct, national test items	56 (0.5)	58 (0.7)	

Source Martin et al. (2012, 2016) and Mullis et al. (2020). Standard errors in parentheses

The TIMSS 2019 curriculum questionnaire was administered online with the suggestion to draw "on the expertise of curriculum specialists and educators"[2] to judge whether an item falls within the curriculum, but without documentation of the procedures used or measures for reliability checks (Martin et al., 2020).

The test items in TIMSS usually assigned one score point for a correct answer, but in each cycle a few test items could assign up to two score points, with one point assigned for a partially correct answer to the test item. The number of score points that were assigned to the test items solved by students differed slightly between cycles. By contrast, according to responses to the curriculum questionnaires, there were considerable variations in the number of score points assigned to test items in a cycle that were considered within the national curriculum. In general, fewer test items were considered to be covered by the national curricula in science than in mathematics, and there was greater variation in the number of test items not reported as covered by the national curricula in science than mathematics—both between countries and within countries between cycles. As around one-third of the test items were released and replaced with new items in each cycle, some variation should be expected. However, the swapping out of test items does not seem able to explain the changes seen in differences between the total number of test-item score points and score points for test items in the national curriculum. For example, Table 4.2 shows a fall from 149 score points for test items in the Norwegian national science curriculum in the 2011 cycle to 116 score points in the 2015 cycle before increasing again to 146 score points in 2019. Whether the 2015 rise in the number of excluded test items due to previous items being replaced by test items outside the curriculum, should be reflected in the TCMA for the 2019 cycle, which would still include items introduced in 2015. As the tested grade level changed in Norway from fourth to fifth grade in the 2015 cycle, the Norwegian drop seems to be for some unaccountable reason. There were similar patterns in science for Sweden (score points 2011: 152, 2015: 107, 2019: 131) and mathematics for Denmark (score points 2011: 179, 2015: 146, 2019: 176). It seems implausible that these fluctuations were caused by the swapping out of test items alone as all test items were based on the TIMSS Assessment Framework, which only underwent minor revisions between 2011, 2015, and 2019 (Mullis & Martin, 2013, 2017).

Turning our attention to the consequences of the measurement of students' ability by the TIMSS assessments, the 2019 analyses showed no clear and significant differences between the countries' average scale scores based on all test items and scale scores based only on country-specific test items. Observed differences ranged from zero points (Finland, both for mathematics and science, Norway for mathematics) to four points on the TIMSS scale (Sweden for mathematics). The rest of the analyses showed differences of one (Denmark for mathematics) or two points on the TIMSS scale.

---

[2] Quotation from the introduction to the TIMSS 2019 online curriculum questionnaire (https://tim ssandpirls.bc.edu/timss2019/questionnaires/index.html).

A significant difference was seen in the percentage of correct test-item score points between the 2011 and 2015 cycles. The difference varied between countries, subjects, and cycles, but there was a clear pattern showing a higher difference in the years when a country had marked relatively many of the test items as being outside the curriculum. The TCMAs showed no clear patterns of coherence between the TIMSS Assessment Framework, and the test items considered part of the individual country's curriculum for each cycle. In some cases, there appeared to be a strong and stable connection with little variation in the number of excluded items (e.g., for mathematics in Finland and Norway) while in other cases, the differences seemed more substantial (e.g., the generally lower number of included items for mathematics in Sweden). Nonetheless, there was variation across cycles, (e.g., for mathematics in Denmark).

4.2.2   Test Items Covered by the Nordic Curricula Over Time

As described above in Sect. 4.2.1, there were relatively large variations in the number of test items considered not covered in the national curricula. Examining these test items (see the respective appendices of the international reports on each TIMSS cycle, Martin et al., 2012, 2016; Mullis et al., 2012a, 2016a, 2020), revealed inconsistencies across the three cycles in all four countries in terms of which test items were counted as covered in the national curricula. In all countries, there are examples of test items that were considered part of the curriculum in one cycle but not covered in the next cycle—and vice versa—with no obvious link to changes to the national curriculum. While some of these differences might be ascribable to adjustments in the respective curricula, as described in Sect. 4.3.1, it must be assumed that others were caused by national changes in interpretations of which test items were within the curriculum.

Hence, the results of the TCMA should be regarded as an indicator of the degree to which the curriculum covered the assessment framework rather than a clear indication of whether or not specific test items were within the curriculum. Further, these inconsistencies demonstrate that rather than a statement of objective truth, such measures are based on subjective judgments. However, we agree with Wagner and Hastedt's (2022) conclusion that TIMSS provides a relatively accurate measurement of student performance also in relation to the national curriculum. This certainly seems to be the case in the Nordic countries, with comparisons between the 2019 assessments of achievement based on all test items and achievement based solely on those test items included in the national curriculum showing only minor differences—likewise when looking at cases where relatively many test-items were considered outside the curriculum. Nonetheless, there was an unexplained variance between the three cycles of TIMSS in the number of excluded items within countries.

4.3 TIMSS and the Intended Curriculum

The intended curriculum in a school or country changes over time, governed by changes in national legislation and, to the extent that local adaptation is permitted, changes in municipal or school-level curriculum frameworks. The following section describes broad overall changes in the national curricula for Denmark, Finland, Norway, and Sweden based on the reporting of the respective curricula in the Encyclopedia entries in the international reports for each cycle of TIMSS, including a few comments on the Icelandic curriculum based on national curriculum documents.

4.3.1 The Nordic Curricula and Changes Over Time

A simple overview of curricular development in the Nordic countries is provided in Table 4.3. For the 2019 cycle of TIMSS, all countries reported that a national curriculum was in place with some possibilities for local (municipal or school-level) adaptations.

All the Nordic countries had presented the national curriculum in a format describing the intended content for each subject across a range of grade levels. However, these grade-specific curriculum objectives were not aligned with the grade levels at which TIMSS was conducted, with the exception of science in Denmark and the 2011 TIMSS cycle in Norway. Thus, the curricula described what students should have achieved one (Finland) or two (Denmark for mathematics, Norway, and Sweden for both subjects) grade levels later than those measured in TIMSS. Consequently, it is difficult to ascertain precise learning objectives for the point at which TIMSS was administered as no particular order was stipulated for the implementation of various elements of the curriculum.

Table 4.3 shows that the Nordic countries have changed their respective curricula at different times relative to the last three TIMSS cycles. In Norway, there was no reform of the curriculum between the three cycles, but while TIMSS 2011 assessed fourth grade students, later cycles assessed students in fifth grade. As such, it might be expected that students in the 2015 and 2019 cycles would have covered more of the curriculum objectives than students in the 2011 cycle. A reform of the Swedish curriculum was implemented after the 2011 cycle of TIMSS with some later minor revisions, and reforms were enacted in Finland and Denmark between the 2015 and 2019 cycles. In Iceland, a new curriculum came into force in 2014 following a revision of the school act in 2008.

Looking across the curriculum changes implemented in the four countries that participated in TIMSS, in all cases, a revision seemed to be curriculum-wide, implementing changes in both mathematics and science. According to the TIMSS Encyclopedias (Kelly et al., 2020; Mullis et al., 2012b, 2016b), the curricula all outlined a number of broad motivations and an overall purpose for the subject, as well as underlining the importance of nurturing students' engagement and self-confidence

Table 4.3 Overview of curriculum revisions in the years 2011 to 2019

	TIMSS cycle	Curriculum in force	Grades covered, mathematics	Grades covered, science
Denmark	2019	2015	Grade 1–3, 4–6, 7–9	Grade 1–2, 3–4, 5–6
	2015	2003	Grade 1–3, 4–6, 7–9	Grade 1–2, 3–4, 5–6
	2011	2003	Grade 1–3, 4–6, 7–9	Grade 1–2, 3–4, 5–6
Finland	2019	2016	Grade 1–2, 3–6, 7–9	Grade 1–2, 3–6, 7–9
	2015	2004	Grade 1–2, 3–5, 6–9	Grade 1–5
	2011	2004	Grade 1–2, 3–5, 6–9	Grade 1–5
Norway	2019	2006	Grade 1–2, 3–4, 5–7, 8–10	Grade 1–4, 5–7
	2015	2006	Grade 1–2, 3–4, 5–7, 8–10	Grade 1–4, 5–7
	2011	2006	Grade 1–2, 3–4, 5–7, 8–10	Grade 1–4, 5–7
Sweden	2019	2011, with minor revisions in 2017/18	Grade 1–3, 4–6, 7–9	Grade 1–3, 4–6, 7–9
	2015	2011	Grade 1–3, 4–6, 7–9	Grade 1–3, 4–6, 7–9
	2011	1994	Goals to be met by grade 5	Goals to be met by grade 5
Iceland	2019	2014	Grade 1–4, 5–7	Grade 1–4, 5–7
	2015	2014	Grade 1–4, 5–7	Grade 1–4, 5–7
	2011	2007	Grade 1–4, 5–7	Grade 1–4, 5–7

in the subject. Further, they stipulated general goals, which was also the case for the Icelandic curriculum (Icelandic Ministry of Education Science & Culture, 2014). Revisions shifted descriptions towards more specific learning goals that students should reach by the end of the grade levels covered.

4.4 Students' Opportunity to Learn—The Implemented Curriculum According to Teachers

Students' opportunity to learn (OTL) specific content areas (Scheerens, 2017) was measured using the teacher questionnaire, which asked teachers whether they had taught specific content in previous years, during the current school year leading up to TIMSS, or whether they expected to cover it later. Consequently, teachers had no opportunity to indicate that a given content area was not part of the national curriculum. The National Research Coordinators (NRC) also reported on whether or not the various content areas were part of the national curriculum at the fourth-grade level, with the same content areas included in the curriculum and teacher questionnaires. Appendix 1 shows the items used, whether an area was considered within the national curriculum, and provides an overview of the content coverage as reported by teachers for the last three cycles of TIMSS (2011, 2015, and 2019). As the TIMSS Assessment Framework has changed over time, the corresponding questions to both teachers and NRCs concerning content coverage also changed slightly between cycles.

4.4.1   Content Coverage in Mathematics in the Intended and Implemented Curriculum

For fourth grade assessment in TIMSS, mathematics content was divided into three content domains across the three cycles. In 2019, the three content domains were labeled number, measurement and geometry, and data. In each cycle, 50 percent of test items in the assessment relate to the content domain number, while the proportion of test items assessing measurement and geometry decreased from 35 to 30% from 2015 to 2019, with a corresponding increase in test items concerning data from 15 to 20%.

The curriculum questionnaire and teacher questionnaire were used to determine whether topics within each of these three content domains were included in the national (intended) curriculum and covered in class (implemented curriculum), with 17 items in the 2011 and 2015 cycles and 18 items in the 2019 cycle. For example, within the content domain data, teachers and NRCs were asked whether students had or would be expected to have worked with "reading and representing data from tables, pictographs, bar graphs, line graphs, and pie charts". As this item illustrates, the questions addressing whether a topic had been covered could address multiple sub-topics. An overview of the items is provided in Appendix 1, which also illustrates that there were minor changes in phrasing between cycles.

Based on responses from NRCs, there were fluctuations between countries and cycles concerning the number of topics that were expected to have been covered, with no clear link to changes to the curriculum or the number of topics considered outside the respective curriculum (Sect. 4.2.2). Norway, where there were no major revisions to the national curriculum between cycles, provides an illustrative case. Despite the switch from assessing grade four students to grade five students between the 2011 and 2015 cycles, there was no change in the number of topics that were expected to have been covered. This was followed by an increase in the number of excluded topics in 2019.

Integers and the four basic arithmetic operations seemed to comprise the core content within mathematics, considered part of the curriculum by NRCs across all cycles and countries except for Denmark in 2011, and taught to between 96 and 100% of students across all cycles. All other topics within mathematics seemed to be less central, and with large variations. When a topic was not considered within the curriculum by the NRC, it tended to have lower coverage by teachers, but all topics had been presented to some students, however, in cases where the NRC indicated that a topic was included in the curriculum during one cycle but not the next, or vice versa, teachers' responses regarding their implementation of this topic in classroom teaching did not reflect such fluctuations.

4.4.2 Content Coverage in Science in the Intended and Implemented Curriculum

Fourth grade science content was likewise divided into three content domains: life science, physical science, and earth science. These domains were covered by 45%, 35%, and 20% of the test items respectively. There was an increase in the number of items in the curriculum questionnaire and teacher questionnaire concerning whether various topics were included in the curriculum and covered in class, rising from 20 items in 2011 to 26 items in 2019. There were likewise changes in how each topic was worded. For instance, in 2011, the topic of fossils was referred to as "fossils of animals and plants (age, location, formation)"; in 2015, the word "understanding" was part of the question on the topic; and in 2019, it was rephrased as "fossils and what they can tell us about past conditions on Earth", thus shifting the focus.

Among teacher responses, the content domain of life science had the highest coverage, with an apparent correlation between the number of students encountering each topic and whether the topic was considered part of the national curriculum. However, the content domain of life science differed from the content domain of number within mathematics in the sense that no topic was covered by teachers to the same extent as the central topics within number. The topic with the highest coverage within life science was "physical and behavioral characteristics of living things and major groups of living things (e.g., mammals, birds, insects, flowering plants)" with 93% coverage by the end of fourth grade among students in Finland in 2015 and 91% in 2019. However, the vast majority of topics had less than 90% coverage.

In general, more topics were considered outside the curriculum in science than in mathematics by the NRCs, with between 25 and 69% of topics considered not included in the country's national curriculum in a given cycle. The exception was Sweden, which only reported one topic that was not included in each of the three cycles. Of the 26 topics identified in the 2019 questionnaires, 7 topics were considered not included in the Danish curriculum, and 17 not included in the Finnish and Norwegian curricula.

Compared to mathematics, larger variations were found in teachers' coverage of topics within science, which was to be expected given the NRCs' indication that fewer topics were included in the respective national curricula. However, teachers reported covering all topics to a limited extent, with some correlation as to whether or not NRCs indicated that the topic was included in the national curriculum.

4.5   The Attained Curriculum—Student Learning and Its Relationship with the Implemented Curriculum

In TIMSS, the attained curriculum is measured by an overall score and scores within each of the three content domains in both mathematics and science, based on all test items. This implies that the measure of attained curriculum covers more than the intended and implemented curriculum, as described above in Sect. 4.2. The following section describes how this measure relates to the implemented curriculum.

4.5.1   The Attained Curriculum in Mathematics

Figure 4.1 shows the development in teachers' reported content coverage within the three different mathematical content domains across the three cycles, as well as student attainment. As described in the previous sections, there were some variations in content coverage within countries. It is notable that, at the national level, an increase or decrease in content coverage in one content domain was shadowed by similar changes in other content domains. One notable exception can be observed in Finland, where coverage of the data domain decreased sharply from 2015 to 2019 following changes in the curriculum.

Average student scores within the different content domains generally fluctuated in line with the overall score, with minor variations between cycles. The large increase in scores seen in Norway from 2011 to 2015 can be attributed to the shift in the target population from grade four to grade five students between the two cycles. However, it is noteworthy that this shift did not seem to lead to major changes in the reported content coverage in geometry, which decreased with each cycle from 2011 to 2019. While the content domains numbers and data saw minor increases in coverage from 2011 to 2015, for data, this increase was not significantly greater than the increase between the 2015 and 2019 cycles, suggesting a general trend to focus more on topics related to processing and interpreting data rather than a direct result of an additional year of teaching.

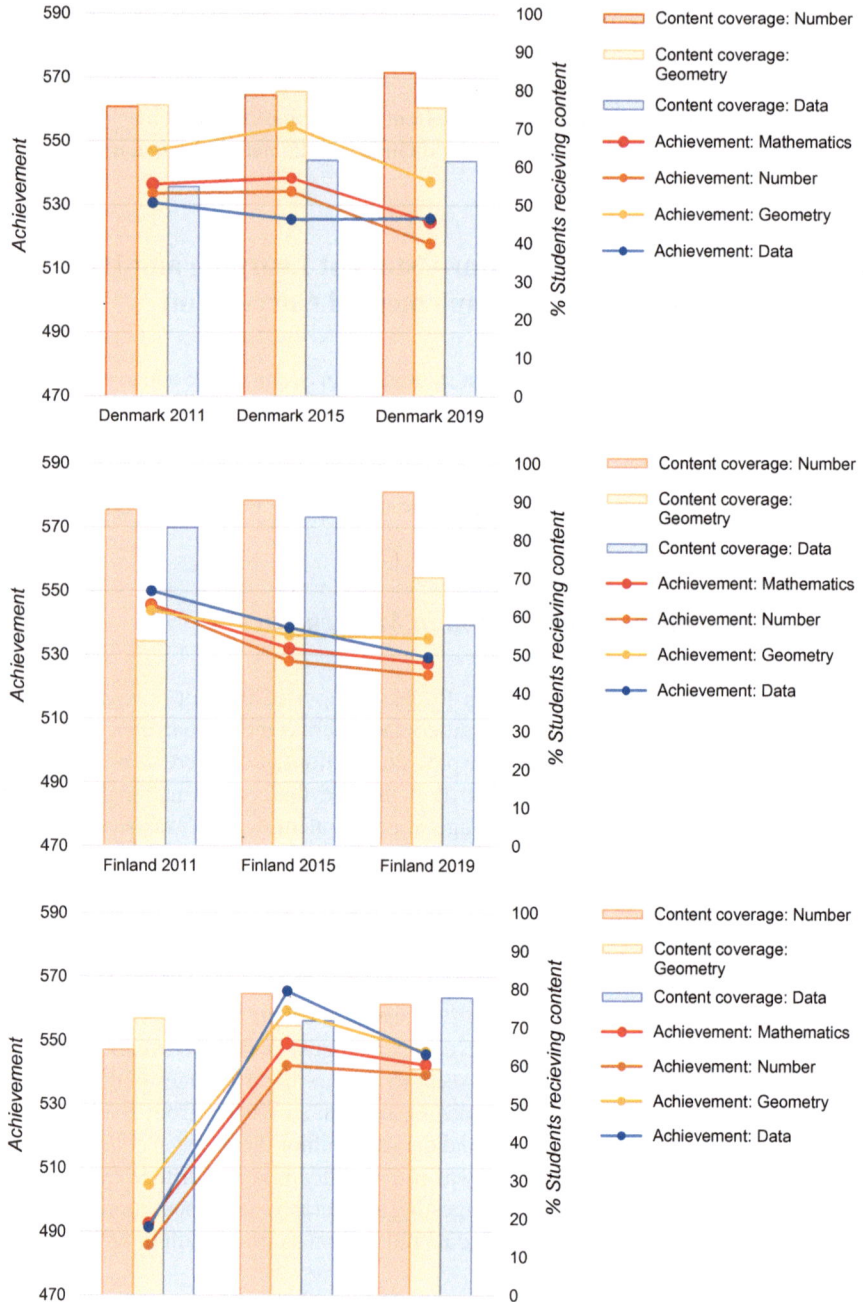

Fig. 4.1 Bar graphs showing achievement in mathematics content domains for the years 2011 to 2019 for all countries. *Note* One bar graph per country.

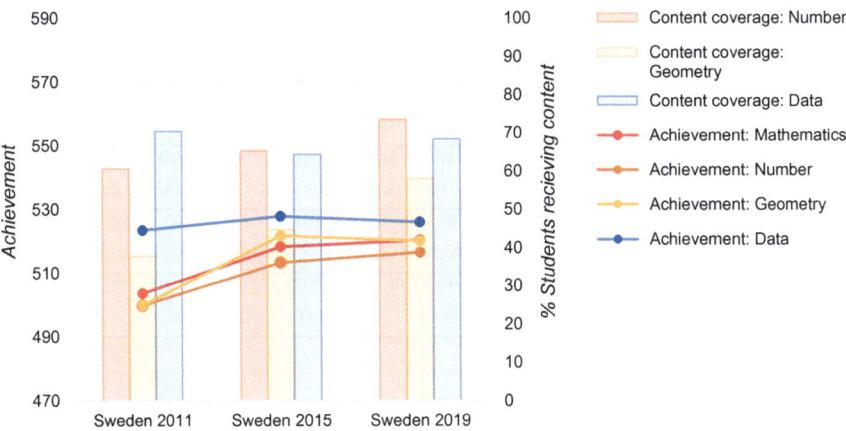

Fig. 4.1 (continued)

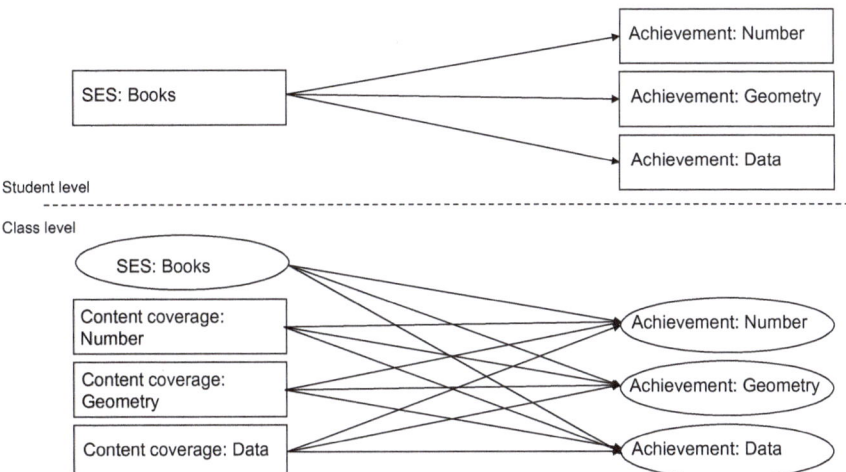

Fig. 4.2 Structural equation model of the relationships between content coverage and scores. *Note* Content coverage = percentage of students taught topics.

The data used to produce Fig. 4.1 are presented in greater detail in the table in Appendix 2, divided into teachers with and without specialization as a mathematics teacher. This revealed differences in both content coverage and achievement scores, with specialized mathematics teachers in general covering slightly more of the curriculum and their students achieving slightly higher scores on average, although the differences go in both directions. It should be noted that not all countries require specialized mathematics training to teach the subject.

Appendix 3 presents the results of a structural equation model (SEM) of the relationship between content coverage in the three content domains and student scores in these domains for all three cycles, as described in Chap. 3. The model, as shown in Fig 4.2, controls for student socioeconomic background (SES) using the student-reported number of books in the home, as students in general are found to profit differently from teaching quality based on their SES level (Atlay et al., 2019).

Some clear relationships are seen in certain countries in some years. However, although the number of significant predictions (all being positive) exceeded what would be expected by chance alone, no clear pattern emerged in the analyses in terms of which content domain predicts scores or across countries.

4.5.2 The Attained Curriculum in Science

Content coverage for the three content domains life science, physical science, and earth science fluctuated slightly between cycles and between countries, with physical science having a lower degree of coverage than the other two content domains in all countries. A higher proportion of missing data was observed for content coverage in science than for mathematics in all countries and cycles, which means that the results should be interpreted with some reservations, especially in relation to Denmark where the rate of missing data exceeded 50 percent for all content domains in each cycle. Appendix 4 provides similar content to Fig. 4.1 for the subject science.

Examining student scores in the different content domains as well as the overall score revealed similar patterns to those found for mathematics. The scores followed similar trends within countries across cycles, with a minor exception being an increase in the Danish students' average score in earth science between 2015 and 2019 while scores in the other domains decreased between these two cycles.

Dividing student scores and content coverage between teachers with and without subject specialization in science, as presented in Appendix 2, some variations were found in both content domain scores and content coverage—both within countries between cycles and within cycles across countries. Once again, it should be noted that subject specialization in science is not a requirement in all countries.

The SEM analyses assessing whether coverage in science content domains predicted student scores within these domains are presented in Appendix 4. These analyses revealed a slight deviation from the patterns previously described for mathematics. Content coverage in physical science significantly predicted student scores in all three content domains for Denmark in 2011 and Finland in 2015, as well as predicting Swedish students' scores in physical science and earth science in 2019. However, a negative correlation was found between content coverage in earth science and scores in physical and earth science in Finland in 2019. Thus, while there seemed to be some significant correlations between content coverage and achievement within and across content domains, there were no consistent patterns across countries or cycles.

4.6 Conclusion

The starting point for this chapter is the didactical expectation that there is (and should be) a connection between the intended, implemented, and achieved curriculum. The analyses presented here do not corroborate the conclusion from Wagner and Hastedt (2022) that TIMSS can be used to draw inferences about the performance of education systems, including those of the Nordic countries, due to the study's coverage of national curricula. Instead, the analyses highlight difficulties in measuring and describing the intended, implemented, and achieved curriculum.

The chapter shows that the measures of implemented curriculum in TIMSS correlate with the measures of achieved curriculum in terms of students' scores in mathematics and science (for further analyses, see Chap. 6). However, the results also illustrate divergence between the different measures of intended and implemented curriculum, which are less reliable than one might hope. Based on the analyses, a range of possible explanations for the low reliability of the measures can be identified.

The first of these explanations relates to differences in how content is defined across the different measures. National curricula are described in general terms with content covering broad areas that students should be taught—often within a time span covering a longer period than the grade where outcomes are measured in TIMSS. As a result, decisions as to whether or not a TIMSS measure of intended, implemented, or achieved curriculum falls within the national curriculum must be based on subjective judgment by the NRC.

Secondly, the terminology that teachers use in their daily work may differ from the terms used in the questionnaires developed to measure teachers' implementation of the curriculum. This can once again introduce reliability issues by requiring interpretations of questions by teachers, which might be complicated further by some questions being ambiguous and hence, force teachers to make a choice about content coverage if only parts of the content are covered. As outlined in the description of curriculum development processes, local adaptations of national curricula can likewise muddy the waters.

Thirdly, whether or not a test item concerns a topic covered by the national curriculum is based on an assessment of whether or not it is included in the curriculum's description of learning objectives and an estimate of the point at which it is taught, given that the curricula generally cover a period extending beyond the TIMSS assessment. For example, there are test items covering the order of operations when using parentheses in mathematics. While not directly mentioned in the Danish curriculum, this is something students would be expected to have learned by the end of the period covered by the curriculum, which ends with sixth grade, but it is difficult to determine more precisely whether this is a topic that will have been covered at the time of the TIMSS assessment. One possible solution might be to consult commonly used textbooks to determine whether the use of parentheses is generally introduced.

Fourthly, and especially relevant when measuring the intended curriculum, some measures are reported for the whole country by a single person, whether the NRC or someone delegated the task by the NRC. Thus, the uncertainties outlined in our previous point will have more serious implications for reliability than the measures conducted at the teacher or student level, where the samples are much larger. These variations in measures may be explained by a change of NRC or in the staff reporting on behalf of the NRC with other content knowledge or other preferences related questionnaire answering.

As indicated throughout this chapter, changes between cycles in the TIMSS Assessment Framework have led to changes in the phrasing of questions collecting information on the intended and implemented curriculum. One limitation of the analyses presented here is that they do not consider how such changes are implemented in the national translations of the teacher questionnaire or whether the translations have been revised without any changes in the international source.

The identified changes in the TIMSS Assessment Framework and its measures of content coverage at different levels are to be expected as they reflect developments in national curricula and teaching practices. As such, we conclude that this framework is well-suited to measuring attained curriculum in the forms of Nordic students' achievement in mathematics and science and to monitoring changes over time despite reliability issues in measuring development in specific areas within the curriculum.

Appendices

Appendix 1 OTL Descriptives from TIMSS 2011, 2015, and 2019

OTL Descriptives 2011—Students

2011

Number	Denmark			Finland		
	Topic inclusion in curriculum[a]	Proportion of students with teacher reporting topic taught in or before G4 (%)	Average achievement by content domain[b]	Topic inclusion in curriculum[a]	Proportion of students with teacher reporting topic taught in or before G4 (%)	Average achievement by content domain[b]
(a) Concepts of whole numbers, including place value and ordering	Not included in the curriculum through grade 4	99	534	All or almost all students	99	545
(b) Adding, subtracting, multiplying, and/or dividing with whole numbers	All or almost all students	98		All or almost all students	100	
(c) Concepts of fractions (fractions as parts of a whole or of a collection, or as a location on a number line; comparing and ordering fractions)	Not included in the curriculum through grade 4	88		All or almost all students	93	
(d) Adding and subtracting with fractions	Not included in the curriculum through grade 4	31		Not included in the curriculum through grade 4	86	

(continued)

(continued)

2011		Denmark			Finland		
		Topic inclusion in curriculum[a]	Proportion of students with teacher reporting topic taught in or before G4 (%)	Average achievement by content domain[b]	Topic inclusion in curriculum[a]	Proportion of students with teacher reporting topic taught in or before G4 (%)	Average achievement by content domain[b]
	(e) Concepts of decimals, including place value and ordering	Not included in the curriculum through grade 4	68		Not included in the curriculum through grade 4	75	
	(f) Adding and subtracting with decimals	Not included in the curriculum through grade 4	75		Not included in the curriculum through grade 4	70	
	(g) Number sentences (finding the missing number, modeling simple situations with number sentences)	Not included in the curriculum through grade 4	58		All or almost all students	79	
	(h) Number patterns (extending number patterns and finding missing terms)	All or almost all students	84		All or almost all students	95	
Measurement & Geometry	(a) Lines: measuring, estimating length of; parallel and perpendicular lines	All or almost all students	92	548	All or almost all students	80	543

(continued)

(continued)

2011

	Denmark			Finland		
	Topic inclusion in curriculum[a]	Proportion of students with teacher reporting topic taught in or before G4 (%)	Average achievement by content domain[b]	Topic inclusion in curriculum[a]	Proportion of students with teacher reporting topic taught in or before G4 (%)	Average achievement by content domain[b]
(b) Comparing and drawing angles	All or almost all students	73		All or almost all students	68	
(c) Using informal coordinate systems to locate points in a plane (e.g., in square B4)	All or almost all students	81		All or almost all students	62	
(d) Elementary properties of common geometric shapes	All or almost all students	90		All or almost all students	84	
(e) Reflections and rotations	All or almost all students	76		All or almost all students	43	
(f) Relationships between two-dimensional and three-dimensional shapes	Not included in the curriculum through grade 4	41		Not included in the curriculum through grade 4	17	
(g) Finding and estimating areas, perimeters and volumes	Not included in the curriculum through grade 4	74		Not included in the curriculum through grade 4	28	

(continued)

(continued)

(continued)

2011		Denmark			Finland		
		Topic inclusion in curriculum[a]	Proportion of students with teacher reporting topic taught in or before G4 (%)	Average achievement by content domain[b]	Topic inclusion in curriculum[a]	Proportion of students with teacher reporting topic taught in or before G4 (%)	Average achievement by content domain[b]
Data	(a) Reading data from tables, pictographs, bar graphs, or pie charts	All or almost all students	68	532	All or almost all students	92	551
	(b) Drawing conclusions from data displays	All or almost all students	54		All or almost all students	76	
	(c) Displaying data using tables, pictographs, and bar graphs	All or almost all students	40		All or almost all students	80	
Life Science	(a) Major body structures and their functions in humans and other organisms (plants and animals)	All or almost all students	72	530	All or almost all students	66	574
	(b) Life cycles and reproduction in plants and animals	All or almost all students	56		Not included in the curriculum through grade 4	73	
	(c) Physical features, behavior, and survival of organisms living in different environments	All or almost all students	67		All or almost all students	71	

(continued)

2011		Denmark			Finland		
		Topic inclusion in curriculum[a]	Proportion of students with teacher reporting topic taught in or before G4 (%)	Average achievement by content domain[b]	Topic inclusion in curriculum[a]	Proportion of students with teacher reporting topic taught in or before G4 (%)	Average achievement by content domain[b]
	(d) Relationships in a given community (e.g., simple food chains, predator/prey relationships)	All or almost all students	75		Not included in the curriculum through grade 4	86	
	(e) Changes in environments (effects of human activity, pollution and its prevention)	All or almost all students	51		Not included in the curriculum through grade 4	72	
	(f) Human health (e.g., transmission/ prevention of communicable diseases, signs of health/illness, diet, exercise)	All or almost all students	56		All or almost all students	71	
Physical Science	(a) States of matter (solids, liquids, gases) and differences in their physical properties (shape, volume), including changes in state of matter by heating and cooling	Not included in the curriculum through grade 4	47	526	All or almost all students	71	568

(continued)

(continued)

2011

	Denmark			Finland		
	Topic inclusion in curriculum[a]	Proportion of students with teacher reporting topic taught in or before G4 (%)	Average achievement by content domain[b]	Topic inclusion in curriculum[a]	Proportion of students with teacher reporting topic taught in or before G4 (%)	Average achievement by content domain[b]
(b) Classification of objects/materials based on physical properties (e.g., weight/mass, volume, magnetic attraction)	Not included in the curriculum through grade 4	41		Not included in the curriculum through grade 4	36	
(c) Forming and separating mixtures	Not included in the curriculum through grade 4	26		Not included in the curriculum through grade 4	13	
(d) Familiar changes in materials (e.g., decaying, burning, rusting, cooking)	All or almost all students	29		All or almost all students	35	
(e) Common energy sources/forms and their practical uses (e.g., the Sun, electricity, water, wind)	All or almost all students	79		Not included in the curriculum through grade 4	64	
(f) Light (e.g., sources, behavior)	All or almost all students	51		All or almost all students	72	

(continued)

(continued)

2011

	Denmark			Finland		
	Topic inclusion in curriculum[a]	Proportion of students with teacher reporting topic taught in or before G4 (%)	Average achievement by content domain[b]	Topic inclusion in curriculum[a]	Proportion of students with teacher reporting topic taught in or before G4 (%)	Average achievement by content domain[b]
(g) Electrical circuits and properties of magnets	Not included in the curriculum through grade 4	59		All or almost all students	52	
(h) Forces that cause objects to move (e.g., gravity, push/pull forces)	Not included in the curriculum through grade 4	54		Not included in the curriculum through grade 4	26	
Earth Science						
(a) Water on Earth (location, types, and movement) and air (composition, proof of its existence, uses)	All or almost all students	58	527	All or almost all students	54	566
(b) Common features of Earth's landscape (e.g., mountains, plains, rivers, deserts) and relationship to human use (e.g., farming, irrigation, land development)	All or almost all students	54		Not included in the curriculum through grade 4	52	
(c) Weather conditions from day to day or over the seasons	All or almost all students	83		All or almost all students	82	

(continued)

(continued)

2011	Denmark			Finland		
	Topic inclusion in curriculum[a]	Proportion of students with teacher reporting topic taught in or before G4 (%)	Average achievement by content domain[b]	Topic inclusion in curriculum[a]	Proportion of students with teacher reporting topic taught in or before G4 (%)	Average achievement by content domain[b]
(d) Fossils of animals and plants (age, location, formation)	All or almost all students	24		Not included in the curriculum through grade 4	7	
(e) Earth's solar system (planets, Sun, moon)	All or almost all students	63		All or almost all students	63	
(f) Day, night, and shadows due to Earth's rotation and its relationship to the Sun	Not included in the curriculum through grade 4	65		All or almost all students	63	

2011

Number		Norway			Sweden		
		Topic inclusion in curriculum[a]	Proportion of students with teacher reporting topic taught in or before G4 (%)	Average achievement by content domain[b]	Topic inclusion in curriculum[a]	Proportion of students with teacher reporting topic taught in or before G4 (%)	Average achievement by content domain[b]
	(a) Concepts of whole numbers, including place value and ordering	All or almost all students	100	488	All or almost all students	100	500
	(b) Adding, subtracting, multiplying, and/or dividing with whole numbers	All or almost all students	98		All or almost all students	99	
	(c) Concepts of fractions (fractions as parts of a whole or of a collection, or as a location on a number line; comparing and ordering fractions)	All or almost all students	50		All or almost all students	41	
	(d) Adding and subtracting with fractions	Not included in the curriculum through grade 4	18		Not included in the curriculum through grade 4	18	
	(e) Concepts of decimals, including place value and ordering	All or almost all students	43		All or almost all students	20	

(continued)

(continued)

2011

	Norway			Sweden		
	Topic inclusion in curriculum[a]	Proportion of students with teacher reporting topic taught in or before G4 (%)	Average achievement by content domain[b]	Topic inclusion in curriculum[a]	Proportion of students with teacher reporting topic taught in or before G4 (%)	Average achievement by content domain[b]
(f) Adding and subtracting with decimals	Not included in the curriculum through grade 4	27		All or almost all students	16	
(g) Number sentences (finding the missing number, modeling simple situations with number sentences)	Not included in the curriculum through grade 4	75		All or almost all students	93	
(h) Number patterns (extending number patterns and finding missing terms)	All or almost all students	90		All or almost all students	94	
Measurement and Geometry (a) Lines: measuring, estimating length of; parallel and perpendicular lines	All or almost all students	74	507	All or almost all students	59	500
(b) Comparing and drawing angles	All or almost all students	50		All or almost all students	21	

(continued)

(continued)

2011

	Norway			Sweden		
	Topic inclusion in curriculum[a]	Proportion of students with teacher reporting topic taught in or before G4 (%)	Average achievement by content domain[b]	Topic inclusion in curriculum[a]	Proportion of students with teacher reporting topic taught in or before G4 (%)	Average achievement by content domain[b]
(c) Using informal coordinate systems to locate points in a plane (e.g., in square B4)	All or almost all students	77		All or almost all students	27	
(d) Elementary properties of common geometric shapes	All or almost all students	92		All or almost all students	85	
(e) Reflections and rotations	All or almost all students	87		Not included in the curriculum through grade 4	12	
(f) Relationships between two-dimensional and three-dimensional shapes	All or almost all students	51		All or almost all students	17	
(g) Finding and estimating areas, perimeters and volumes	All or almost all students	81		All or almost all students	37	
Data (a) Reading data from tables, pictographs, bar graphs, or pie charts	Not included in the curriculum through grade 4	76	494	All or almost all students	79	523

(continued)

(continued)

2011	Norway			Sweden		
	Topic inclusion in curriculum[a]	Proportion of students with teacher reporting topic taught in or before G4 (%)	Average achievement by content domain[b]	Topic inclusion in curriculum[a]	Proportion of students with teacher reporting topic taught in or before G4 (%)	Average achievement by content domain[b]
(b) Drawing conclusions from data displays	All or almost all students	48		All or almost all students	75	
(c) Displaying data using tables, pictographs, and bar graphs	All or almost all students	56		All or almost all students	70	
Life Science (a) Major body structures and their functions in humans and other organisms (plants and animals)	All or almost all students	69	496	All or almost all students	45	534
(b) Life cycles and reproduction in plants and animals	All or almost all students	68		All or almost all students	67	
(c) Physical features, behavior, and survival of organisms living in different environments	Not included in the curriculum through grade 4	41		All or almost all students	66	
(d) Relationships in a given community (e.g., simple food chains, predator—prey relationships)	Not included in the curriculum through grade 4	83		All or almost all students	62	

(continued)

(continued)

2011		Norway			Sweden		
		Topic inclusion in curriculum[a]	Proportion of students with teacher reporting topic taught in or before G4 (%)	Average achievement by content domain[b]	Topic inclusion in curriculum[a]	Proportion of students with teacher reporting topic taught in or before G4 (%)	Average achievement by content domain[b]
	(e) Changes in environments (effects of human activity, pollution and its prevention)	All or almost all students	74		All or almost all students	59	
	(f) Human health (e.g., transmission/prevention of communicable diseases, signs of health/illness, diet, exercise)	All or almost all students	74		All or almost all students	37	
Physical Science	(a) States of matter (solids, liquids, gases) and differences in their physical properties (shape, volume), including changes in state of matter by heating and cooling	All or almost all students	53	482	All or almost all students	55	528
	(b) Classification of objects/materials based on physical properties (e.g., weight/mass, volume, magnetic attraction)	All or almost all students	17		All or almost all students	27	
	(c) Forming and separating mixtures	Not included in the curriculum through grade 4	9		All or almost all students	32	

(continued)

(continued)

2011

	Norway			Sweden		
	Topic inclusion in curriculum[a]	Proportion of students with teacher reporting topic taught in or before G4 (%)	Average achievement by content domain[b]	Topic inclusion in curriculum[a]	Proportion of students with teacher reporting topic taught in or before G4 (%)	Average achievement by content domain[b]
(d) Familiar changes in materials (e.g., decaying, burning, rusting, cooking)	All or almost all students	46		All or almost all students	29	
(e) Common energy sources/forms and their practical uses (e.g., the Sun, electricity, water, wind)	Not included in the curriculum through grade 4	61		All or almost all students	52	
(f) Light (e.g., sources, behavior)	All or almost all students	41		All or almost all students	17	
(g) Electrical circuits and properties of magnets	Not included in the curriculum through grade 4	10		All or almost all students	33	
(h) Forces that cause objects to move (e.g., gravity, push/pull forces)	Not included in the curriculum through grade 4	22		All or almost all students	14	

(continued)

(continued)

2011

	Norway			Sweden		
	Topic inclusion in curriculum[a]	Proportion of students with teacher reporting topic taught in or before G4 (%)	Average achievement by content domain[b]	Topic inclusion in curriculum[a]	Proportion of students with teacher reporting topic taught in or before G4 (%)	Average achievement by content domain[b]
Earth Science						
(a) Water on Earth (location, types, and movement) and air (composition, proof of its existence, uses)	Not included in the curriculum through grade 4	65	506	All or almost all students	57	538
(b) Common features of Earth's landscape (e.g., mountains, plains, rivers, deserts) and relationship to human use (e.g., farming, irrigation, land development)	Not included in the curriculum through grade 4	65		All or almost all students	60	
(c) Weather conditions from day to day or over the seasons	All or almost all students	90		All or almost all students	84	
(d) Fossils of animals and plants (age, location, formation)	All or almost all students	72		Not included in the curriculum through grade 4	56	
(e) Earth's solar system (planets, Sun, moon)	All or almost all students	95		All or almost all students	88	

(continued)

(continued)

2011	Norway			Sweden		
	Topic inclusion in curriculum[a]	Proportion of students with teacher reporting topic taught in or before G4 (%)	Average achievement by content domain[b]	Topic inclusion in curriculum[a]	Proportion of students with teacher reporting topic taught in or before G4 (%)	Average achievement by content domain[b]
(f) Day, night, and shadows due to Earth's rotation and its relationship to the Sun	All or almost all students	81		All or almost all students	82	

NB [a] The Curriculum questionnaire for TIMSS 2011 asked NRCs: "According to the national mathematics curriculum, what proportion of grade 4 students should have been taught each of the following topics or skills by the end of grade 4?" and "According to the national science curriculum, what proportion of grade 4 students should have been taught each of the following topics or skills by the end of grade 4?"
[b] (Martin et al., 2012; Mullis et al., 2012a, 2012b)

OTL Descriptives 2015—Students

	Denmark			Finland		
	Topic inclusion in curriculum[a]	Proportion of students with teacher reporting topic taught in or before G4 (%)	Average achievement by content domain[b]	Topic inclusion in curriculum[a]	Proportion of students with teacher reporting topic taught in or before G4 (%)	Average achievement by content domain[b]
Number						
(a) Concepts of whole numbers, including place value and ordering	Topics Taught to All or Almost All Students	100	535	Topics Taught to All or Almost All Students	97	532
(b) Adding, subtracting, multiplying, and/or dividing with whole numbers	Topics Taught to All or Almost All Students	99		Topics Taught to All or Almost All Students	100	
(c) Concepts of multiples and factors; odd and even numbers	Topics Taught to All or Almost All Students	87		Topics Taught to All or Almost All Students	94	
(d) Concepts of fractions (fractions as parts of a whole or of a collection, or as a location on a number line)	Topics Taught to All or Almost All Students	88		Topics Taught to All or Almost All Students	92	
(e) Adding and subtracting with fractions, comparing and ordering fractions	Topics Taught to All or Almost All Students	46		Topics Taught to All or Almost All Students	86	

(continued)

(continued)

	Denmark			Finland		
	Topic inclusion in curriculum[a]	Proportion of students with teacher reporting topic taught in or before G4 (%)	Average achievement by content domain[b]	Topic inclusion in curriculum[a]	Proportion of students with teacher reporting topic taught in or before G4 (%)	Average achievement by content domain[b]
(f) Concepts of decimals, including place value and ordering, adding and subtracting with decimals	Topics Taught to All or Almost All Students	81		Topics Taught to All or Almost All Students	75	
(g) Number sentences (finding the missing number, modeling simple situations with number sentences)	Topics Taught to All or Almost All Students	44		Topics Taught to All or Almost All Students	78	
(h) Number patterns (extending number patterns and finding missing terms)	Topics Taught to All or Almost All Students	81		Topics Taught to All or Almost All Students	98	
Measurement and Geometry (a) Lines: measuring, estimating length of; parallel and perpendicular lines	Topics Taught to All or Almost All Students	95	555	Topics Taught to All or Almost All Students	80	539
(b) Comparing and drawing angles	Topics Taught to All or Almost All Students	80		Topics Taught to All or Almost All Students	66	

(continued)

(continued)

	Denmark			Finland		
	Topic inclusion in curriculum[a]	Proportion of students with teacher reporting topic taught in or before G4 (%)	Average achievement by content domain[b]	Topic inclusion in curriculum[a]	Proportion of students with teacher reporting topic taught in or before G4 (%)	Average achievement by content domain[b]
(c) Using informal coordinate systems to locate points in a plane (e.g., in square B4)	Topics Taught to All or Almost All Students	85		Topics Taught to All or Almost All Students	65	
(d) Elementary properties of common geometric shapes	Topics Taught to All or Almost All Students	96		Topics Taught to All or Almost All Students	89	
(e) Reflections and rotations	Topics Taught to Only the More Able Students (Top Track)	75		Topics Taught to All or Almost All Students	43	
(f) Relationships between two-dimensional and three-dimensional shapes	Topics Taught to All or Almost All Students	46		Not Included in the Curriculum Through Grade 4	25	

(continued)

(continued)

		Denmark			Finland		
		Topic inclusion in curriculum[a]	Proportion of students with teacher reporting topic taught in or before G4 (%)	Average achievement by content domain[b]	Topic inclusion in curriculum[a]	Proportion of students with teacher reporting topic taught in or before G4 (%)	Average achievement by content domain[b]
	(g) Finding and estimating areas, perimeters, and volumes	Topics Taught to All or Almost All Students	78		Topics Taught to All or Almost All Students	38	
Data	(a) Reading and representing data from tables, pictographs, bar graphs, or pie charts	Topics Taught to All or Almost All Students	72	526	Topics Taught to All or Almost All Students	92	542
	(b) Drawing conclusions from data displays	Topics Taught to All or Almost All Students	51		Topics Taught to All or Almost All Students	80	
Life Science	(a) Characteristics of living things and the major groups of living things (e.g., mammals, birds, insects, flowering plants)	Topics Taught to All or Almost All Students	79	534	Topics Taught to All or Almost All Students	93	556
	(b) Major body structures and their functions in humans, other animals, and plants	Topics Taught to All or Almost All Students	75		Topics Taught to All or Almost All Students	63	

(continued)

(continued)

	Denmark			Finland		
	Topic inclusion in curriculum[a]	Proportion of students with teacher reporting topic taught in or before G4 (%)	Average achievement by content domain[b]	Topic inclusion in curriculum[a]	Proportion of students with teacher reporting topic taught in or before G4 (%)	Average achievement by content domain[b]
(c) Life cycles of common plants and animals (e.g., humans, butterflies, frogs, flowering plants)	Topics Taught to All or Almost All Students	63		Not Included in the Curriculum Through Grade 4	83	
(d) Understanding that some characteristics are inherited and some are the result of the environment	Not Included in the Curriculum Through Grade 4	32		Not Included in the Curriculum Through Grade 4	33	
(e) How physical features and behaviors help living things survive in their environments	Topics Taught to All or Almost All Students	54		Topics Taught to All or Almost All Students	72	
(f) Relationships in communities and ecosystems (e.g., simple food chains, predator–prey relationships, human impacts on the environment)	Topics Taught to All or Almost All Students	73		Not Included in the Curriculum Through Grade 4	91	

(continued)

(continued)

	Denmark			Finland		
	Topic inclusion in curriculum[a]	Proportion of students with teacher reporting topic taught in or before G4 (%)	Average achievement by content domain[b]	Topic inclusion in curriculum[a]	Proportion of students with teacher reporting topic taught in or before G4 (%)	Average achievement by content domain[b]
(g) Human health (transmission and prevention of diseases, symptoms of health and illness, importance of a healthy diet and exercise)	Topics Taught to All or Almost All Students	64		Topics Taught to All or Almost All Students	75	
Physical Science (a) States of matter (solid, liquid, gas) and properties of the states of matter (volume, shape); how the state of matter changes by heating or cooling	Topics Taught to All or Almost All Students	51	516	Topics Taught to All or Almost All Students	70	547
(b) Classifying materials based on physical properties (e.g., weight/mass, volume, conducting heat, conducting electricity, magnetic attraction)	Topics Taught to All or Almost All Students	38		Not Included in the Curriculum Through Grade 4	39	
(c) Mixtures and how to separate a mixture into its components (e.g., sifting, filtering, evaporation, using a magnet)	Topics Taught to All or Almost All Students	23		Not Included in the Curriculum Through Grade 4	17	

(continued)

(continued)

	Denmark			Finland		
	Topic inclusion in curriculum[a]	Proportion of students with teacher reporting topic taught in or before G4 (%)	Average achievement by content domain[b]	Topic inclusion in curriculum[a]	Proportion of students with teacher reporting topic taught in or before G4 (%)	Average achievement by content domain[b]
(d) Chemical changes in everyday life (e.g., decaying, burning, rusting, cooking)	Topics Taught to All or Almost All Students	23		Topics Taught to All or Almost All Students	25	
(e) Common sources of energy (e.g., the Sun, electricity, wind) and uses of energy (heating and cooling homes, providing light)	Topics Taught to Only the More Able Students (Top Track)	73		Not Included in the Curriculum Through Grade 4	66	
(f) Light and sound in everyday life (e.g., understanding shadows and reflection, understanding that vibrating objects make sound)	Topics Taught to All or Almost All Students	38		Topics Taught to All or Almost All Students	71	
(g) Electricity and simple circuits (e.g., identifying materials that are conductors, recognizing that electricity can be changed to light or sound, knowing that a circuit must be complete to work correctly)	Topics Taught to All or Almost All Students	53		Topics Taught to All or Almost All Students	54	

(continued)

(continued)

	Denmark			Finland		
	Topic inclusion in curriculum[a]	Proportion of students with teacher reporting topic taught in or before G4 (%)	Average achievement by content domain[b]	Topic inclusion in curriculum[a]	Proportion of students with teacher reporting topic taught in or before G4 (%)	Average achievement by content domain[b]
(h) Properties of magnets (e.g., knowing that like poles repel and opposite poles attract, recognizing that magnets can attract some objects)	Topics Taught to All or Almost All Students	59		Topics Taught to All or Almost All Students	44	
(i) Forces that cause objects to move (e.g., gravity, pushing/pulling)	Topics Taught to All or Almost All Students	44		Not Included in the Curriculum Through Grade 4	40	
Earth Science (a) Common features of the Earth's landscape (e.g., mountains, plains, deserts, rivers, oceans) and their relationship to human use (farming, irrigation, land development)	Topics Taught to All or Almost All Students	53	531	Not Included in the Curriculum Through Grade 4	58	560
(b) Where water is found on the Earth and how it moves in and out of the air (e.g., evaporation, rainfall, cloud formation, dew formation)	Topics Taught to All or Almost All Students	67		Topics Taught to All or Almost All Students	78	

(continued)

(continued)

	Denmark			Finland		
	Topic inclusion in curriculum[a]	Proportion of students with teacher reporting topic taught in or before G4 (%)	Average achievement by content domain[b]	Topic inclusion in curriculum[a]	Proportion of students with teacher reporting topic taught in or before G4 (%)	Average achievement by content domain[b]
(c) Understanding that weather can change from day to day, from season to season, and by geographic location	Topics Taught to All or Almost All Students	64		Topics Taught to All or Almost All Students	68	
(d) Understanding what fossils are and what they can tell us about past conditions on Earth	Topics Taught to Only the More Able Students (Top Track)	28		Not Included in the Curriculum Through Grade 4	20	
(e) Objects in the solar system (the Sun, the Earth, the Moon, and other planets) and their movements (the Earth and other planets revolve around the Sun, the Moon revolves around the Earth)	Topics Taught to All or Almost All Students	76		Topics Taught to All or Almost All Students	76	
(f) Understanding how day and night result from the Earth's rotation on its axis and how the Earth's rotation results in changing shadows throughout the day	Topics Taught to All or Almost All Students	72		Topics Taught to All or Almost All Students	78	

(continued)

(continued)

	Denmark			Finland		
	Topic inclusion in curriculum[a]	Proportion of students with teacher reporting topic taught in or before G4 (%)	Average achievement by content domain[b]	Topic inclusion in curriculum[a]	Proportion of students with teacher reporting topic taught in or before G4 (%)	Average achievement by content domain[b]
(g) Understanding how seasons are related to the Earth's annual movement around the Sun	Topics Taught to All or Almost All Students	72		Not Included in the Curriculum Through Grade 4	77	

		Norway			Sweden		
		Topic inclusion in curriculum[a]	Proportion of students with teacher reporting topic taught in or before G4 (%)	Average achievement by content domain[b]	Topic inclusion in curriculum[a]	Proportion of students with teacher reporting topic taught in or before G4 (%)	Average achievement by content domain[b]
Number	(a) Concepts of whole numbers, including place value and ordering	Topics Taught to All or Almost All Students	99	542	Topics Taught to All or Almost All Students	100	514
	(b) Adding, subtracting, multiplying, and/or dividing with whole numbers	Topics Taught to All or Almost All Students	100		Topics Taught to All or Almost All Students	98	
	(c) Concepts of multiples and factors; odd and even numbers	Topics Taught to All or Almost All Students	67		Topics Taught to All or Almost All Students	82	
	(d) Concepts of fractions (fractions as parts of a whole or of a collection, or as a location on a number line)	Topics Taught to All or Almost All Students	78		Topics Taught to All or Almost All Students	42	
	(e) Adding and subtracting with fractions, comparing and ordering fractions	Not Included in the Curriculum Through Grade 4	53		Topics Taught to All or Almost All Students	28	

(continued)

(continued)

	Norway			Sweden		
	Topic inclusion in curriculum[a]	Proportion of students with teacher reporting topic taught in or before G4 (%)	Average achievement by content domain[b]	Topic inclusion in curriculum[a]	Proportion of students with teacher reporting topic taught in or before G4 (%)	Average achievement by content domain[b]
(f) Concepts of decimals, including place value and ordering, adding and subtracting with decimals	Topics Taught to All or Almost All Students	82		Topics Taught to All or Almost All Students	20	
(g) Number sentences (finding the missing number, modeling simple situations with number sentences)	Topics Taught to All or Almost All Students	65		Topics Taught to All or Almost All Students	65	
(h) Number patterns (extending number patterns and finding missing terms)	Topics Taught to All or Almost All Students	80		Topics Taught to All or Almost All Students	86	
Measurement and Geometry (a) Lines: measuring, estimating length of; parallel and perpendicular lines	Not Included in the Curriculum Through Grade 4	84	559	Topics Taught to All or Almost All Students	59	523

(continued)

(continued)

	Norway			Sweden		
	Topic inclusion in curriculum[a]	Proportion of students with teacher reporting topic taught in or before G4 (%)	Average achievement by content domain[b]	Topic inclusion in curriculum[a]	Proportion of students with teacher reporting topic taught in or before G4 (%)	Average achievement by content domain[b]
(b) Comparing and drawing angles	Not Included in the Curriculum Through Grade 4	77		Topics Taught to All or Almost All Students	41	
(c) Using informal coordinate systems to locate points in a plane (e.g., in square B4)	Topics Taught to All or Almost All Students	69		Topics Taught to All or Almost All Students	22	
(d) Elementary properties of common geometric shapes	Topics Taught to All or Almost All Students	91		Topics Taught to All or Almost All Students	89	
(e) Reflections and rotations	Topics Taught to All or Almost All Students	71		Topics Taught to All or Almost All Students	22	

(continued)

(continued)

		Norway			Sweden		
		Topic inclusion in curriculum[a]	Proportion of students with teacher reporting topic taught in or before G4 (%)	Average achievement by content domain[b]	Topic inclusion in curriculum[a]	Proportion of students with teacher reporting topic taught in or before G4 (%)	Average achievement by content domain[b]
	(f) Relationships between two-dimensional and three-dimensional shapes	Not Included in the Curriculum Through Grade 4	31		Topics Taught to All or Almost All Students	30	
	(g) Finding and estimating areas, perimeters, and volumes	Topics Taught to All or Almost All Students	61		Topics Taught to All or Almost All Students	42	
Data	(a) Reading and representing data from tables, pictographs, bar graphs, or pie charts	Topics Taught to All or Almost All Students	84	566	Topics Taught to All or Almost All Students	63	529
	(b) Drawing conclusions from data displays	Topics Taught to All or Almost All Students	61		Topics Taught to All or Almost All Students	62	

(continued)

(continued)

		Norway			Sweden		
		Topic inclusion in curriculum[a]	Proportion of students with teacher reporting topic taught in or before G4 (%)	Average achievement by content domain[b]	Topic inclusion in curriculum[a]	Proportion of students with teacher reporting topic taught in or before G4 (%)	Average achievement by content domain[b]
Life Science	(a) Characteristics of living things and the major groups of living things (e.g., mammals, birds, insects, flowering plants)	Not Included in the Curriculum Through Grade 4	82	546	Topics Taught to All or Almost All Students	79	540
	(b) Major body structures and their functions in humans, other animals, and plants	Topics Taught to All or Almost All Students	85		Topics Taught to All or Almost All Students	45	
	(c) Life cycles of common plants and animals (e.g., humans, butterflies, frogs, flowering plants)	Topics Taught to All or Almost All Students	63		Topics Taught to All or Almost All Students	70	
	(d) Understanding that some characteristics are inherited and some are the result of the environment	Not Included in the Curriculum Through Grade 4	45		Topics Taught to All or Almost All Students	29	

(continued)

(continued)

	Norway			Sweden		
	Topic inclusion in curriculum[a]	Proportion of students with teacher reporting topic taught in or before G4 (%)	Average achievement by content domain[b]	Topic inclusion in curriculum[a]	Proportion of students with teacher reporting topic taught in or before G4 (%)	Average achievement by content domain[b]
(e) How physical features and behaviors help living things survive in their environments	Not Included in the Curriculum Through Grade 4	42		Topics Taught to All or Almost All Students	62	
(f) Relationships in communities and ecosystems (e.g., simple food chains, predator–prey relationships, human impacts on the environment)	Topics Taught to All or Almost All Students	54		Topics Taught to All or Almost All Students	74	
(g) Human health (transmission and prevention of diseases, symptoms of health and illness, importance of a healthy diet and exercise)	Topics Taught to All or Almost All Students	64		Topics Taught to All or Almost All Students	38	
Physical Science (a) States of matter (solid, liquid, gas) and properties of the states of matter (volume, shape); how the state of matter changes by heating or cooling	Not Included in the Curriculum Through Grade 4	54	522	Topics Taught to All or Almost All Students	84	534

(continued)

(continued)

	Norway			Sweden		
	Topic inclusion in curriculum[a]	Proportion of students with teacher reporting topic taught in or before G4 (%)	Average achievement by content domain[b]	Topic inclusion in curriculum[a]	Proportion of students with teacher reporting topic taught in or before G4 (%)	Average achievement by content domain[b]
(b) Classifying materials based on physical properties (e.g., weight/mass, volume, conducting heat, conducting electricity, magnetic attraction)	Not Included in the Curriculum Through Grade 4	29		Topics Taught to All or Almost All Students	35	
(c) Mixtures and how to separate a mixture into its components (e.g., sifting, filtering, evaporation, using a magnet)	Not Included in the Curriculum Through Grade 4	44		Topics Taught to All or Almost All Students	60	
(d) Chemical changes in everyday life (e.g., decaying, burning, rusting, cooking)	Not Included in the Curriculum Through Grade 4	51		Topics Taught to All or Almost All Students	42	
(e) Common sources of energy (e.g., the Sun, electricity, wind) and uses of energy (heating and cooling homes, providing light)	Not Included in the Curriculum Through Grade 4	48		Topics Taught to All or Almost All Students	56	

(continued)

(continued)

Topic	Norway			Sweden		
	Topic inclusion in curriculum[a]	Proportion of students with teacher reporting topic taught in or before G4 (%)	Average achievement by content domain[b]	Topic inclusion in curriculum[a]	Proportion of students with teacher reporting topic taught in or before G4 (%)	Average achievement by content domain[b]
(f) Light and sound in everyday life (e.g., understanding shadows and reflection, understanding that vibrating objects make sound)	Not Included in the Curriculum Through Grade 4	28		Topics Taught to All or Almost All Students	26	
(g) Electricity and simple circuits (e.g., identifying materials that are conductors, recognizing that electricity can be changed to light or sound, knowing that a circuit must be complete to work correctly)	Not Included in the Curriculum Through Grade 4	17		Topics Taught to All or Almost All Students	41	
(h) Properties of magnets (e.g., knowing that like poles repel and opposite poles attract, recognizing that magnets can attract some objects)	Not Included in the Curriculum Through Grade 4	63		Topics Taught to All or Almost All Students	26	
(i) Forces that cause objects to move (e.g., gravity, pushing/pulling)	Not Included in the Curriculum Through Grade 4	42		Topics Taught to All or Almost All Students	37	

(continued)

(continued)

		Norway			Sweden		
		Topic inclusion in curriculum[a]	Proportion of students with teacher reporting topic taught in or before G4 (%)	Average achievement by content domain[b]	Topic inclusion in curriculum[a]	Proportion of students with teacher reporting topic taught in or before G4 (%)	Average achievement by content domain[b]
Earth Science	(a) Common features of the Earth's landscape (e.g., mountains, plains, deserts, rivers, oceans) and their relationship to human use (farming, irrigation, land development)	Not Included in the Curriculum Through Grade 4	67	549	Topics Taught to All or Almost All Students	53	552
	(b) Where water is found on the Earth and how it moves in and out of the air (e.g., evaporation, rainfall, cloud formation, dew formation)	Not Included in the Curriculum Through Grade 4	57		Topics Taught to All or Almost All Students	89	
	(c) Understanding that weather can change from day to day, from season to season, and by geographic location	Topics Taught to All or Almost All Students	70		Topics Taught to All or Almost All Students	63	
	(d) Understanding what fossils are and what they can tell us about past conditions on Earth	Topics Taught to All or Almost All Students	68		Not Included in the Curriculum Through Grade 4	55	

(continued)

(continued)

	Norway			Sweden		
	Topic inclusion in curriculum[a]	Proportion of students with teacher reporting topic taught in or before G4 (%)	Average achievement by content domain[b]	Topic inclusion in curriculum[a]	Proportion of students with teacher reporting topic taught in or before G4 (%)	Average achievement by content domain[b]
(e) Objects in the solar system (the Sun, the Earth, the Moon, and other planets) and their movements (the Earth and other planets revolve around the Sun, the Moon revolves around the Earth)	Topics Taught to All or Almost All Students	71		Topics Taught to All or Almost All Students	77	
(f) Understanding how day and night result from the Earth's rotation on its axis and how the Earth's rotation results in changing shadows throughout the day	Not Included in the Curriculum Through Grade 4	71		Topics Taught to All or Almost All Students	75	
(g) Understanding how seasons are related to the Earth's annual movement around the Sun	Not Included in the Curriculum Through Grade 4	68		Topics Taught to All or Almost All Students	76	

NB [a] The Curriculum questionnaire for TIMSS 2015 asked NRCs: "According to the national mathematics curriculum, what proportion of grade 4 students should have been taught each of the following topics or skills by the end of grade 4?" and "According to the national science curriculum, what proportion of grade 4 students should have been taught each of the following topics or skills by the end of grade 4?"
[b] (Martin et al., 2016; Mullis et al., 2016)

OTL Descriptives 2019—Students

2019		Denmark			Finland		
		Topic inclusion in curriculum[a]	Proportion of students with teacher reporting topic taught in or before G4 (%)	Average achievement by content domain[b]	Topic inclusion in curriculum[a]	Proportion of students with teacher reporting topic taught in or before G4 (%)	Average achievement by content domain[b]
Number	(a) Concepts of whole numbers, including place value and ordering	All or almost all the students	98	518	All or almost all the students	98	528
	(b) Adding, subtracting, multiplying, and dividing with whole numbers	All or almost all the students	97		All or almost all the students	100	
	(c) Concepts of multiples and factors; odd and even numbers	All or almost all the students	82		All or almost all the students	93	
	(d) Number sentences (finding the missing number, representing problem situations with number sentences)	All or almost all the students	81		Only the more able students	74	
	(e) Number patterns (extending number patterns and finding missing terms)	All or almost all the students	84		All or almost all the students	97	
	(f) Concepts of fractions, including representing, comparing and ordering, adding and subtracting simple fractions	All or almost all the students	73		All or almost all the students	98	

(continued)

(continued)

2019	Denmark			Finland		
	Topic inclusion in curriculum[a]	Proportion of students with teacher reporting topic taught in or before G4 (%)	Average achievement by content domain[b]	Topic inclusion in curriculum[a]	Proportion of students with teacher reporting topic taught in or before G4 (%)	Average achievement by content domain[b]
(g) Concepts of decimals, including place value and ordering, adding and subtracting with decimals	All or almost all the students	77		All or almost all the students	94	
Measurement and Geometry			536			538
(a) Solving problems involving length, including measuring and estimating	All or almost all the students	92		All or almost all the students	86	
(b) Solving problems involving mass, volume, and time	All or almost all the students	57		All or almost all the students	66	
(c) Finding and estimating perimeter, area, and volume	All or almost all the students	86		All or almost all the students	52	
(d) Parallel and perpendicular lines	All or almost all the students	88		All or almost all the students	89	
(e) Comparing and drawing angles	All or almost all the students	83		All or almost all the students	71	
(f) Elementary properties of common geometric shapes	All or almost all the students	95		All or almost all the students	88	

(continued)

(continued)

2019	Denmark			Finland		
	Topic inclusion in curriculum[a]	Proportion of students with teacher reporting topic taught in or before G4 (%)	Average achievement by content domain[b]	Topic inclusion in curriculum[a]	Proportion of students with teacher reporting topic taught in or before G4 (%)	Average achievement by content domain[b]
(g) Three-dimensional shapes, including relationships with their two-dimensional representations	All or almost all the students	33		Not included in the curriculum through Grade 4	32	
Data (a) Reading and representing data from tables, pictographs, bar graphs, line graphs, and pie charts	All or almost all the students	70	525	All or almost all the students	72	534
(b) Organizing and representing data to help answer questions	All or almost all the students	63		All or almost all the students	48	
(c) Drawing conclusions from data displays	All or almost all the students	54		All or almost all the students	54	
Life Science (a) Physical and behavioral characteristics of living things and major groups of living things (e.g., mammals, birds, insects, flowering plants)	All or almost all the students	88	526	All or almost all the students	91	558

(continued)

(continued)

2019	Denmark			Finland		
	Topic inclusion in curriculum[a]	Proportion of students with teacher reporting topic taught in or before G4 (%)	Average achievement by content domain[b]	Topic inclusion in curriculum[a]	Proportion of students with teacher reporting topic taught in or before G4 (%)	Average achievement by content domain[b]
(b) Major body structures and their functions in humans, other animals, and plants	All or almost all the students	78		All or almost all the students	60	
(c) Life cycles of common plants and animals (e.g., flowering plants, butterflies, frogs)	All or almost all the students	79		All or almost all the students	79	
(d) Characteristics of plants and animals that are inherited	Not included in the curriculum through Grade 4	37		Not included in the curriculum through Grade 4	17	
(e) Interactions between organisms and their environments (e.g., physical features and behaviors that help living things survive in their environments)	All or almost all the students	56		Not included in the curriculum through Grade 4	68	
(f) Relationships in ecosystems (e.g., simple food chains, predator–prey relationships, competition)	All or almost all the students	85		All or almost all the students	89	

(continued)

(continued)

2019	Denmark			Finland		
	Topic inclusion in curriculum[a]	Proportion of students with teacher reporting topic taught in or before G4 (%)	Average achievement by content domain[b]	Topic inclusion in curriculum[a]	Proportion of students with teacher reporting topic taught in or before G4 (%)	Average achievement by content domain[b]
(g) Human health (transmission and prevention of diseases, everyday behaviors that promote good health)	All or almost all the students	64		All or almost all the students	83	
Physical science						
(a) States of matter (solid, liquid, gas) and their properties (volume, mass, volume, state of matter, shape)	All or almost all the students	72	507	All or almost all the students	80	544
(b) Classifying materials based on physical properties (e.g., weight/mass, volume, state of matter, conductivity of heat or electricity)	All or almost all the students	39		Not included in the curriculum through Grade 4	38	
(c) Mixtures, including methods for separating a mixture into its components (e.g., sifting, filtering, evaporation, using a magnet)	All or almost all the students	27		Not included in the curriculum through Grade 4	18	

(continued)

(continued)

2019	Denmark			Finland		
	Topic inclusion in curriculum[a]	Proportion of students with teacher reporting topic taught in or before G4 (%)	Average achievement by content domain[b]	Topic inclusion in curriculum[a]	Proportion of students with teacher reporting topic taught in or before G4 (%)	Average achievement by content domain[b]
(d) Properties of magnets (e.g., like poles repel and opposite poles attract, magnets can attract some objects)	All or almost all the students	44		Not included in the curriculum through Grade 4	36	
(e) Physical changes in everyday life (e.g., changes of state, dissolving)	Not included in the curriculum through Grade 4	37		Not included in the curriculum through Grade 4	50	
(f) Chemical changes in everyday life (e.g., decaying, burning, rusting, cooking)	Not included in the curriculum through Grade 4	35		Not included in the curriculum through Grade 4	36	
(g) Common sources of energy (e.g., the Sun, wind, oil) and uses of energy (heating and cooling homes, providing light)	Not included in the curriculum through Grade 4	67		Not included in the curriculum through Grade 4	63	

(continued)

(continued)

2019	Denmark			Finland		
	Topic inclusion in curriculum[a]	Proportion of students with teacher reporting topic taught in or before G4 (%)	Average achievement by content domain[b]	Topic inclusion in curriculum[a]	Proportion of students with teacher reporting topic taught in or before G4 (%)	Average achievement by content domain[b]
(h) Light and sound in everyday life (e.g., shadows and reflections, vibrating objects make sound)	All or almost all the students	44		All or almost all the students	63	
(i) Heat transfer (e.g., energy flows from a hot object to a colder object)	Not included in the curriculum through Grade 4	41		Not included in the curriculum through Grade 4	34	
(j) Electricity and simple electrical circuits (e.g., a circuit must be complete to work correctly)	All or almost all the students	63		Not included in the curriculum through Grade 4	39	
(k) Forces that cause objects to move (e.g., gravity, pushing/pulling) or change their motion (e.g., friction)	Not included in the curriculum through Grade 4	49		Not included in the curriculum through Grade 4	34	

(continued)

(continued)

2019	Denmark			Finland		
	Topic inclusion in curriculum[a]	Proportion of students with teacher reporting topic taught in or before G4 (%)	Average achievement by content domain[b]	Topic inclusion in curriculum[a]	Proportion of students with teacher reporting topic taught in or before G4 (%)	Average achievement by content domain[b]
(l) Simple machines (e.g., levers, pulleys, wheels, ramps) that help make motion easier	Not included in the curriculum through Grade 4	39		All or almost all the students	56	
Earth Science						
(a) Physical makeup of Earth's surface (e.g., land and water in unequal proportions, sources of fresh and salt water)	All or almost all the students	64	535	Not included in the curriculum through Grade 4	60	563
(b) Earth's resources used in everyday life (e.g., water, wind, soil, forests, oil, natural gas, minerals)	All or almost all the students	65		Not included in the curriculum through Grade 4	48	
(c) Changes in Earth's surface over time (e.g., mountain building, weathering, erosion)	All or almost all the students	47		Not included in the curriculum through Grade 4	21	

(continued)

(continued)

2019	Denmark			Finland		
	Topic inclusion in curriculum[a]	Proportion of students with teacher reporting topic taught in or before G4 (%)	Average achievement by content domain[b]	Topic inclusion in curriculum[a]	Proportion of students with teacher reporting topic taught in or before G4 (%)	Average achievement by content domain[b]
(d) Fossils and what they can tell us about past conditions on Earth	All or almost all the students	41		Not included in the curriculum through Grade 4	16	
(e) Weather and climate (e.g., daily, seasonal, and locational variations versus long term trends)	All or almost all the students	83		Not included in the curriculum through Grade 4	71	
(f) Objects in the Solar System (the Sun, the Earth, the Moon, and other planets) and their movements	All or almost all the students	84		Not included in the curriculum through Grade 4	70	
(g) Earth's motion and related patterns observed on Earth (e.g., day and night, seasons)	All or almost all the students	86		All or almost all the students	82	

2019		Norway			Sweden		
		Topic inclusion in curriculum[a]	Proportion of students with teacher reporting topic taught in or before G4 (%)	Average achievement by content domain[b]	Topic inclusion in curriculum[a]	Proportion of students with teacher reporting topic taught in or before G4 (%)	Average achievement by content domain[b]
Number	(a) Concepts of whole numbers, including place value and ordering	All or almost all the students	100	540	All or almost all the students	100	517
	(b) Adding, subtracting, multiplying, and dividing with whole numbers	All or almost all the students	96		All or almost all the students	100	
	(c) Concepts of multiples and factors; odd and even numbers	All or almost all the students	67		All or almost all the students	86	
	(d) Number sentences (finding the missing number, representing problem situations with number sentences)	All or almost all the students	66		All or almost all the students	68	
	(e) Number patterns (extending number patterns and finding missing terms)	All or almost all the students	67		All or almost all the students	85	
	(f) Concepts of fractions, including representing, comparing and ordering, adding and subtracting simple fractions	Not included in the curriculum through Grade 4	51		Not included in the curriculum through Grade 4	48	
	(g) Concepts of decimals, including place value and ordering, adding and subtracting with decimals	All or almost all the students	90		Not included in the curriculum through Grade 4	24	

(continued)

(continued)

2019		Norway			Sweden		
		Topic inclusion in curriculum[a]	Proportion of students with teacher reporting topic taught in or before G4 (%)	Average achievement by content domain[b]	Topic inclusion in curriculum[a]	Proportion of students with teacher reporting topic taught in or before G4 (%)	Average achievement by content domain[b]
Measurement and Geometry	(a) Solving problems involving length, including measuring and estimating	All or almost all the students	82	546	All or almost all the students	86	521
	(b) Solving problems involving mass, volume, and time	Not included in the curriculum through Grade 4	37		All or almost all the students	56	
	(c) Finding and estimating perimeter, area, and volume	Not included in the curriculum through Grade 4	75		All or almost all the students	47	

(continued)

(continued)

2019	Norway			Sweden		
	Topic inclusion in curriculum[a]	Proportion of students with teacher reporting topic taught in or before G4 (%)	Average achievement by content domain[b]	Topic inclusion in curriculum[a]	Proportion of students with teacher reporting topic taught in or before G4 (%)	Average achievement by content domain[b]
(d) Parallel and perpendicular lines	Not included in the curriculum through Grade 4	37		Not included in the curriculum through Grade 4	39	
(e) Comparing and drawing angles	Not included in the curriculum through Grade 4	70		Not included in the curriculum through Grade 4	49	
(f) Elementary properties of common geometric shapes	All or almost all the students	87		All or almost all the students	88	
(g) Three-dimensional shapes, including relationships with their two-dimensional representations	Not included in the curriculum through Grade 4	33		All or almost all the students	34	

(continued)

(continued)

2019	Norway			Sweden		
	Topic inclusion in curriculum[a]	Proportion of students with teacher reporting topic taught in or before G4 (%)	Average achievement by content domain[b]	Topic inclusion in curriculum[a]	Proportion of students with teacher reporting topic taught in or before G4 (%)	Average achievement by content domain[b]
Data						
(a) Reading and representing data from tables, pictographs, bar graphs, line graphs, and pie charts	Not included in the curriculum through Grade 4	86	547	Not included in the curriculum through Grade 4	81	527
(b) Organizing and representing data to help answer questions	All or almost all the students	79		All or almost all the students	61	
(c) Drawing conclusions from data displays	Not included in the curriculum through Grade 4	71		All or almost all the students	63	
Life Science						
(a) Physical and behavioral characteristics of living things and major groups of living things (e.g., mammals, birds, insects, flowering plants)	Not included in the curriculum through Grade 4	75	547	All or almost all the students	78	541
(b) Major body structures and their functions in humans, other animals, and plants	All or almost all the students	88		All or almost all the students	52	

(continued)

(continued)

2019	Norway			Sweden		
	Topic inclusion in curriculum[a]	Proportion of students with teacher reporting topic taught in or before G4 (%)	Average achievement by content domain[b]	Topic inclusion in curriculum[a]	Proportion of students with teacher reporting topic taught in or before G4 (%)	Average achievement by content domain[b]
(c) Life cycles of common plants and animals (e.g., flowering plants, butterflies, frogs)	All or almost all the students	64		All or almost all the students	69	
(d) Characteristics of plants and animals that are inherited	Not included in the curriculum through Grade 4	27		All or almost all the students	33	
(e) Interactions between organisms and their environments (e.g., physical features and behaviors that help living things survive in their environments)	Not included in the curriculum through Grade 4	47		All or almost all the students	65	
(f) Relationships in ecosystems (e.g., simple food chains, predator–prey relationships, competition)	All or almost all the students	62		All or almost all the students	80	
(g) Human health (transmission and prevention of diseases, everyday behaviors that promote good health)	All or almost all the students	67		All or almost all the students	38	

(continued)

(continued)

2019	Norway			Sweden		
	Topic inclusion in curriculum[a]	Proportion of students with teacher reporting topic taught in or before G4 (%)	Average achievement by content domain[b]	Topic inclusion in curriculum[a]	Proportion of students with teacher reporting topic taught in or before G4 (%)	Average achievement by content domain[b]
Physical science			525			525
(a) States of matter (solid, liquid, gas) and their properties (volume, shape)	Not included in the curriculum through Grade 4	50		All or almost all the students	85	
(b) Classifying materials based on physical properties (e.g., weight/mass, volume, state of matter, conductivity of heat or electricity)	Not included in the curriculum through Grade 4	21		All or almost all the students	43	
(c) Mixtures, including methods for separating a mixture into its components (e.g., sifting, filtering, evaporation, using a magnet)	Not included in the curriculum through Grade 4	36		All or almost all the students	46	
(d) Properties of magnets (e.g., like poles repel and opposite poles attract, magnets can attract some objects)	Not included in the curriculum through Grade 4	66		All or almost all the students	20	

(continued)

(continued)

2019	Norway			Sweden		
	Topic inclusion in curriculum[a]	Proportion of students with teacher reporting topic taught in or before G4 (%)	Average achievement by content domain[b]	Topic inclusion in curriculum[a]	Proportion of students with teacher reporting topic taught in or before G4 (%)	Average achievement by content domain[b]
(e) Physical changes in everyday life (e.g., changes of state, dissolving)	All or almost all the students	46		All or almost all the students	51	
(f) Chemical changes in everyday life (e.g., decaying, burning, rusting, cooking)	Not included in the curriculum through Grade 4	43		All or almost all the students	43	
(g) Common sources of energy (e.g., the Sun, wind, oil) and uses of energy (heating and cooling homes, providing light)	Not included in the curriculum through Grade 4	57		All or almost all the students	48	
(h) Light and sound in everyday life (e.g., shadows and reflections, vibrating objects make sound)	All or almost all the students	27		All or almost all the students	17	
(i) Heat transfer (e.g., energy flows from a hot object to a colder object)	Not included in the curriculum through Grade 4	19		All or almost all the students	38	

(continued)

(continued)

2019	Norway			Sweden		
	Topic inclusion in curriculum[a]	Proportion of students with teacher reporting topic taught in or before G4 (%)	Average achievement by content domain[b]	Topic inclusion in curriculum[a]	Proportion of students with teacher reporting topic taught in or before G4 (%)	Average achievement by content domain[b]
(j) Electricity and simple electrical circuits (e.g., a circuit must be complete to work correctly)	Not included in the curriculum through Grade 4	18		All or almost all the students	24	
(k) Forces that cause objects to move (e.g., gravity, pushing/pulling) or change their motion (e.g., friction)	Not included in the curriculum through Grade 4	39		All or almost all the students	26	
(l) Simple machines (e.g., levers, pulleys, wheels, ramps) that help make motion easier	Not included in the curriculum through Grade 4	19		All or almost all the students	42	
Earth Science (a) Physical makeup of Earth's surface (e.g., land and water in unequal proportions, sources of fresh and salt water)	Not included in the curriculum through Grade 4	47	547	All or almost all the students	59	547

(continued)

(continued)

2019	Norway			Sweden		
	Topic inclusion in curriculum[a]	Proportion of students with teacher reporting topic taught in or before G4 (%)	Average achievement by content domain[b]	Topic inclusion in curriculum[a]	Proportion of students with teacher reporting topic taught in or before G4 (%)	Average achievement by content domain[b]
(b) Earth's resources used in everyday life (e.g., water, wind, soil, forests, oil, natural gas, minerals)	Not included in the curriculum through Grade 4	59		All or almost all the students	53	
(c) Changes in Earth's surface over time (e.g., mountain building, weathering, erosion)	Not included in the curriculum through Grade 4	51		All or almost all the students	36	
(d) Fossils and what they can tell us about past conditions on Earth	All or almost all the students	63		All or almost all the students	47	
(e) Weather and climate (e.g., daily, seasonal, and locational variations versus long term trends)	All or almost all the students	73		Not included in the curriculum through Grade 4	74	
(f) Objects in the Solar System (the Sun, the Earth, the Moon, and other planets) and their movements	All or almost all the students	84		All or almost all the students	66	

(continued)

(continued)

2019	Norway			Sweden		
	Topic inclusion in curriculum[a]	Proportion of students with teacher reporting topic taught in or before G4 (%)	Average achievement by content domain[b]	Topic inclusion in curriculum[a]	Proportion of students with teacher reporting topic taught in or before G4 (%)	Average achievement by content domain[b]
(g) Earth's motion and related patterns observed on Earth (e.g., day and night, seasons)	Not included in the curriculum through Grade 4	83		All or almost all the students	74	

NB [a] The Curriculum questionnaire for TIMSS 2019 asked NRCs: "According to the national mathematics curriculum, what proportion of grade 4 students should have been taught each of the following topics or skills by the end of grade 4?" and "According to the national science curriculum, what proportion of grade 4 students should have been taught each of the following topics or skills by the end of grade 4?"
[b] (Mullis et al., 2020)

Appendix 2 Mean Differences in Student Achievement

		Ach: Math	Ach: Number	Ach: Geometry	Ach: Data	Content coverage: Number	Content coverage: Geometry	Content coverage: Data
Mathematics specialists	Denmark 2011	539.666	536.640	551.680	534.931	75.020	75.678	53.330
	Denmark 2015	531.438	528.070	546.555	518.626	78.492	76.650	56.707
	Denmark 2019	533.003	517.075	536.250	525.878	83.698	73.014	62.103
	Difference 2015–2011	− 8.228	− 8.570	− 5.125	− 16.305	3.472	0.972	3.377
	Difference 2019–2015	1.565	− 10.995	− 10.305	7.252	5.206	− 3.636	5.396
Non-mathematics specialists	Denmark 2011	536.753	533.826	547.231	530.816	75.633	79.970	50.508
	Denmark 2015	537.925	533.944	554.374	525.000	81.044	83.699	66.727
	Denmark 2019	527.124	519.957	539.076	527.427	86.105	78.739	61.445
	Difference 2015–2011	1.172	0.118	7.143	− 5.816	5.411	3.729	16.219
	Difference 2019–2015	− 10.801	− 13.987	− 15.298	2.427	5.061	− 4.960	− 5.282
Mathematics specialists	Finland 2011	553.171	553.695	551.633	555.398	78.611	58.252	78.056
	Finland 2015	524.652	519.145	528.660	529.106	88.381	60.021	89.545
	Finland 2019	539.645	536.386	545.373	542.756	96.040	64.493	57.717
	Difference 2015–2011	− 28.519	− 34.550	− 22.973	− 26.292	9.770	1.769	11.489
	Difference 2019–2015	14.993	17.241	16.713	13.650	7.659	4.472	− 31.828
Non-mathematics specialists	Finland 2011	545.847	545.740	543.955	551.398	89.555	52.838	83.732
	Finland 2015	532.015	528.376	535.967	538.436	90.845	56.203	86.052

(continued)

(continued)

		Ach: Math	Ach: Number	Ach: Geometry	Ach: Data	Content coverage: Number	Content coverage: Geometry	Content coverage: Data
	Finland 2019	526.688	522.597	533.174	527.969	92.708	70.914	58.460
	Difference 2015–2011	− 13.832	− 17.364	− 7.988	− 12.962	1.290	3.365	2.320
	Difference 2019–2015	− 5.327	− 5.779	− 2.793	− 10.467	1.863	14.711	− 27.592
Mathematics specialists	Norway 2011	496.334	490.566	507.609	495.047	64.139	68.142	70.201
	Norway 2015	552.684	544.794	562.503	570.392	77.950	71.078	76.681
	Norway 2019	546.358	543.305	552.131	552.100	77.590	61.839	77.705
	Difference 2015–2011	56.350	54.228	54.894	75.345	13.811	2.936	6.480
	Difference 2019–2015	− 6.326	− 1.489	− 10.372	− 18.292	− 0.360	− 9.239	1.024
Non-mathematics specialists	Norway 2011	490.421	483.538	483.538	490.356	62.817	73.275	62.904
	Norway 2015	548.927	542.014	558.957	564.939	78.036	70.540	69.900
	Norway 2019	543.070	540.337	545.129	546.239	75.259	56.553	78.300
	Difference 2015–2011	58.506	58.476	75.419	74.583	15.219	− 2.735	6.996
	Difference 2019–2015	− 5.857	− 1.677	− 13.828	− 18.700	− 2.777	− 13.987	8.400
Mathematics specialists	Sweden 2011	501.306	498.133	497.871	522.152	62.050	36.894	66.454
	Sweden 2015	519.396	514.004	523.989	528.909	65.715	42.712	66.355
	Sweden 2019	524.426	520.508	524.799	530.143	74.583	58.849	69.258

(continued)

(continued)

	Ach: Math	Ach: Number	Ach: Geometry	Ach: Data	Content coverage: Number	Content coverage: Geometry	Content coverage: Data
Difference 2015–2011	18.090	15.871	26.118	6.757	3.665	5.818	−0.099
Difference 2019–2015	5.030	6.504	0.810	1.234	8.868	16.137	2.903
Non-mathematics specialists Sweden 2011	510.193	505.495	504.855	527.156	57.967	40.568	75.621
Sweden 2015	514.251	510.542	515.655	528.863	66.172	54.498	51.228
Sweden 2019	515.257	508.482	514.203	518.962	71.984	53.945	82.466
Difference 2015–2011	4.058	5.047	10.800	1.707	8.205	13.930	−24.393
Difference 2019–2015	1.006	−2.060	−1.452	−9.901	5.812	−0.553	31.238
Science specialists Denmark 2011	536.579	533.876	546.881	528.920	69.614	76.920	45.712
Denmark 2015	525.996	522.180	537.839	512.386	76.009	74.488	52.126
Denmark 2019	521.490	515.222	533.757	520.979	83.797	70.912	57.880
Difference 2015–2011	−10.583	−11.696	−9.042	−16.534	6.395	−2.432	6.414
Difference 2019–2015	−4.506	−6.958	−4.082	8.593	7.788	−3.576	5.754
Non-science specialists Denmark 2011	538.387	535.349	549.792	533.382	75.944	78.580	53.677
Denmark 2015	536.499	532.912	553.149	523.639	80.639	81.094	63.168
Denmark 2019	526.712	519.301	538.635	528.096	85.190	77.122	62.542
Difference 2015–2011	−1.888	−2.437	3.357	−9.743	4.695	2.514	9.491

(continued)

(continued)

		Ach: Math	Ach: Number	Ach: Geometry	Ach: Data	Content coverage: Number	Content coverage: Geometry	Content coverage: Data
	Difference 2019–2015	− 9.787	− 13.611	− 14.514	4.457	4.551	− 3.972	− 0.626
Science specialists	Finland 2011	553.587	552.031	551.482	557.107	85.404	50.733	81.624
	Finland 2015	539.813	534.768	541.004	547.842	89.717	50.780	93.970
	Finland 2019	518.167	517.041	523.828	521.694	95.113	73.035	60.274
	Difference 2015–2011	− 13.774	− 17.263	− 10.478	− 9.265	4.313	0.047	12.346
	Difference 2019–2015	− 21.646	− 17.727	− 17.176	− 26.148	5.396	22.255	− 33.696
Non-science specialists	Finland 2011	545.728	545.844	544.064	551.093	88.399	54.036	83.127
	Finland 2015	530.605	526.999	534.913	536.753	90.587	57.166	85.574
	Finland 2019	528.618	524.264	535.202	529.879	92.849	70.453	58.676
	Difference 2015–2011	− 15.123	− 18.845	− 9.151	− 14.340	2.188	3.130	2.447
	Difference 2019–2015	− 1.987	− 2.735	0.289	− 6.874	2.262	13.287	− 26.898
Science Specialists	Norway 2011	484.077	477.362	496.935	483.243	64.655	73.688	82.157
	Norway 2015	553.976	547.298	565.106	570.937	78.262	72.462	82.069
	Norway 2019	552.408	549.394	556.811	558.320	79.336	61.963	81.326
	Difference 2015–2011	69.899	69.936	68.171	87.694	13.607	− 1.226	− 0.088
	Difference 2019–2015	− 1.568	2.096	− 8.295	− 12.617	1.074	− 10.499	− 0.743

(continued)

(continued)

		Ach: Math	Ach: Number	Ach: Geometry	Ach: Data	Content coverage: Number	Content coverage: Geometry	Content coverage: Data
Non-science specialists	Norway 2011	493.742	487.163	505.330	493.470	63.490	72.013	61.066
	Norway 2015	549.312	541.847	559.013	565.682	78.188	70.391	68.919
	Norway 2019	541.859	538.972	545.782	546.013	75.345	58.078	76.467
	Difference 2015–2011	55.570	54.684	53.683	72.212	14.698	− 1.622	7.853
	Difference 2019–2015	− 7.453	− 2.875	− 13.231	− 19.669	− 2.843	− 12.313	7.548
Science specialists	Sweden 2011	502.305	498.864	498.438	523.077	60.995	37.017	66.067
	Sweden 2015	520.707	514.856	525.587	531.049	64.449	42.309	67.624
	Sweden 2019	527.761	524.142	528.984	533.903	74.920	57.029	68.645
	Difference 2015–2011	18.402	15.992	27.149	7.972	3.454	5.292	1.557
	Difference 2019–2015	7.054	9.286	3.397	2.854	10.471	14.720	1.021
Non-science specialists	Sweden 2011	506.768	502.574	502.367	524.687	60.267	40.019	73.727
	Sweden 2015	514.586	511.049	517.301	525.041	68.118	48.553	57.348
	Sweden 2019	513.959	508.039	511.956	518.023	72.555	60.409	76.062
	Difference 2015–2011	7.818	8.475	14.934	0.354	7.851	8.534	− 16.379
	Difference 2019–2015	− 0.627	− 3.010	− 5.345	− 7.018	4.437	11.856	18.714

		Ach: Science	Ach: Life Science	Ach: Physical Science	Ach: Earth Science	Content coverage: Life Science	Content coverage: Physical Science	Content coverage: Earth Science
Mathematics specialists	Denmark 2011	531.021	533.302	528.896	533.302	75.485	53.559	63.560
	Denmark 2015	521.384	529.124	509.614	524.939	56.753	43.259	62.936
	Denmark 2019	520.490	523.827	504.838	536.250	67.471	44.685	65.603
	Difference 2015–2011	− 9.637	− 4.178	− 19.282	− 8.363	− 18.732	− 10.300	− 0.624
	Difference 2019–2015	− 0.894	− 5.297	− 4.776	11.311	10.718	1.426	2.667
Non-mathematics specialists	Denmark 2011	529.326	531.515	527.195	527.427	58.829	45.290	68.896
	Denmark 2015	525.960	533.179	514.243	530.076	65.314	49.615	65.776
	Denmark 2019	524.833	529.159	509.750	537.095	71.050	47.320	62.986
	Difference 2015–2011	− 3.366	1.664	− 12.952	2.649	6.485	4.325	− 3.120
	Difference 2019–2015	− 1.127	− 4.020	− 4.493	7.019	5.736	− 2.295	− 2.790
Mathematics specialists	Finland 2011	576.212	580.262	572.396	567.916	71.213	31.287	45.905
	Finland 2015	545.127	543.965	534.754	552.603	67.633	55.940	60.875
	Finland 2019	557.806	559.214	549.260	570.306	66.365	50.379	55.851
	Difference 2015–2011	− 31.085	− 36.297	− 37.642	− 15.313	− 3.580	24.653	14.970
	Difference 2019–2015	12.679	15.249	14.506	17.703	− 1.268	− 5.561	− 5.024

(continued)

(continued)

		Ach: Science	Ach: Life Science	Ach: Physical Science	Ach: Earth Science	Content coverage: Life Science	Content coverage: Physical Science	Content coverage: Earth Science
Non-mathematics specialists	Finland 2011	570.785	574.620	568.354	566.548	74.785	41.922	54.612
	Finland 2015	549.827	552.288	543.261	556.816	71.932	45.729	64.273
	Finland 2019	549.721	553.675	539.283	557.687	70.218	46.055	53.685
	Difference 2015–2011	−20.958	−22.332	−25.093	−9.732	−2.853	3.807	9.661
	Difference 2019–2015	−0.106	1.387	−3.978	0.871	−1.714	0.326	−10.588
Mathematics specialists	Norway 2011	493.518	495.077	479.741	503.675	62.116	27.221	76.670
	Norway 2015	540.769	549.367	524.299	551.775	60.188	42.882	62.861
	Norway 2019	542.728	549.863	528.162	550.729	63.168	35.292	62.159
	Difference 2015–2011	47.251	54.290	44.558	48.100	−1.928	15.661	−13.809
	Difference 2019–2015	1.959	0.496	3.863	−1.046	2.980	−7.590	−0.702
Non-mathematics specialists	Norway 2011	490.253	491.802	478.996	502.307	65.522	32.511	77.687
	Norway 2015	536.142	544.492	521.046	547.911	66.539	41.532	71.107
	Norway 2019	537.400	546.496	524.447	543.054	58.780	36.186	67.271
	Difference 2015–2011	45.889	52.690	42.050	45.604	1.017	9.021	−6.580
	Difference 2019–2015	1.258	2.004	3.401	−4.857	−7.759	−5.346	−3.836

(continued)

(continued)

		Ach: Science	Ach: Life Science	Ach: Physical Science	Ach: Earth Science	Content coverage: Life Science	Content coverage: Physical Science	Content coverage: Earth Science
Mathematics specialists	Sweden 2011	530.713	531.795	525.673	535.163	57.650	30.765	68.802
	Sweden 2015	541.605	540.506	535.481	552.930	59.585	47.223	70.310
	Sweden 2019	540.823	545.046	528.391	550.573	59.352	41.875	53.506
	Difference 2015–2011	10.892	8.711	9.808	17.767	1.935	16.458	1.508
	Difference 2019–2015	− 0.782	4.540	− 7.090	− 2.357	− 0.233	− 5.348	− 16.804
Non-mathematics specialists	Sweden 2011	540.475	539.131	533.227	545.977	58.065	31.991	67.100
	Sweden 2015	533.441	533.840	526.746	541.831	57.799	33.656	78.632
	Sweden 2019	529.790	534.529	517.861	541.011	60.115	32.774	64.072
	Difference 2015–2011	− 7.034	− 5.291	− 6.481	− 4.146	− 0.266	1.665	11.532
	Difference 2019–2015	− 3.651	0.689	− 8.885	− 0.820	2.316	− 0.882	− 14.560
Science specialists	Denmark 2011	526.807	528.160	524.539	524.539	74.320	44.621	64.634
	Denmark 2015	517.557	526.456	507.180	522.043	50.294	35.732	50.938
	Denmark 2019	514.578	518.516	498.288	528.822	65.090	40.866	63.130
	Difference 2015–2011	-9.250	− 1.704	− 17.359	− 2.496	− 24.026	− 8.889	− 13.696
	Difference 2019–2015	-2.979	− 7.940	− 8.892	6.779	14.796	5.134	12.192

(continued)

(continued)

		Ach: Science	Ach: Life Science	Ach: Physical Science	Ach: Earth Science	Content coverage: Life Science	Content coverage: Physical Science	Content coverage: Earth Science
Non-science specialists	Denmark 2011	530.780	533.227	528.667	529.431	65.102	50.357	64.707
	Denmark 2015	524.948	531.970	512.815	528.682	65.444	48.880	66.449
	Denmark 2019	524.741	528.556	509.618	536.591	69.780	47.528	64.288
	Difference 2015–2011	-5.832	– 1.257	– 15.852	– 0.749	0.342	– 1.477	1.742
	Difference 2019–2015	-0.207	– 3.414	– 3.197	7.909	4.336	– 1.352	– 2.161
Science specialists	Finland 2011	579.493	585.257	576.122	576.333	63.146	37.901	50.173
	Finland 2015	560.065	564.767	555.846	569.759	66.336	46.338	59.372
	Finland 2019	541.578	546.355	531.674	548.432	65.966	45.499	48.558
	Difference 2015–2011	-19.428	– 20.490	– 20.276	– 6.574	3.190	8.437	9.199
	Difference 2019–2015	-18.487	– 18.412	– 24.172	– 21.327	– 0.370	– 0.839	– 10.814
Non-science specialists	Finland 2011	570.405	574.066	567.994	565.505	75.876	40.675	53.800
	Finland 2015	548.535	550.371	541.329	555.115	72.054	46.657	64.725
	Finland 2019	551.414	555.046	541.081	559.890	70.352	46.158	54.128
	Difference 2015–2011	-21.870	– 23.695	– 26.665	– 10.390	– 3.822	5.982	10.925
	Difference 2019–2015	2.879	4.675	– 0.248	4.775	– 1.702	– 0.499	– 10.597

(continued)

(continued)

		Ach: Science	Ach: Life Science	Ach: Physical Science	Ach: Earth Science	Content coverage: Life Science	Content coverage: Physical Science	Content coverage: Earth Science
Science specialists	Norway 2011	482.824	483.750	470.683	492.891	70.897	40.715	80.085
	Norway 2015	541.536	549.208	525.497	553.377	67.767	51.522	66.498
	Norway 2019	549.262	556.438	533.914	555.860	65.263	36.326	62.701
	Difference 2015–2011	58.712	65.458	54.814	60.486	−3.130	10.807	−13.587
	Difference 2019–2015	7.726	7.230	8.417	2.483	−2.504	−15.196	−3.797
Non-science specialists	Norway 2011	492.976	494.724	481.367	505.104	63.363	29.031	77.824
	Norway 2015	536.898	545.725	521.586	548.562	63.309	37.780	69.476
	Norway 2019	536.456	544.866	523.292	543.512	58.723	35.872	68.040
	Difference 2015–2011	43.922	51.001	40.219	43.458	−0.054	8.749	−8.348
	Difference 2019–2015	-0.442	−0.859	1.706	−5.050	−4.586	−1.908	−1.436
Science specialists	Sweden 2011	532.212	533.898	527.525	537.554	59.628	31.561	68.593
	Sweden 2015	543.336	542.818	537.599	555.204	60.272	48.241	69.396
	Sweden 2019	544.463	548.268	532.563	555.095	60.122	40.871	54.034
	Difference 2015–2011	11.124	8.920	10.074	17.650	0.644	16.680	0.803
	Difference 2019–2015	1.127	5.450	−5.036	−0.109	−0.150	−7.370	−15.362

(continued)

(continued)

		Ach: Science	Ach: Life Science	Ach: Physical Science	Ach: Earth Science	Content coverage: Life Science	Content coverage: Physical Science	Content coverage: Earth Science
Non-science specialists	Sweden 2011	536.434	535.026	529.351	540.781	54.296	33.033	69.778
	Sweden 2015	534.601	533.144	527.625	543.749	56.709	34.604	78.038
	Sweden 2019	529.110	534.276	515.958	537.993	56.992	40.485	60.283
	Difference 2015–2011	-1.833	-1.882	-1.726	2.968	2.413	1.571	8.260
	Difference 2019–2015	-5.491	1.132	-11.667	-5.756	0.283	5.881	-17.755

**Appendix 3 The Relationships Between Content Coverage
and Student Achievement at the Within/Student
and Between/Classroom levels**

	2011				2015			
	Denmark	Finland	Norway	Sweden	Denmark	Finland	Norway	Sweden
Within level (Student level)								
Achievement—Data on								
SES(Books)	0.231*	0.204*	0.176*	0.210*	0.263*	0.191*	0.291*	0.252*
Achievement—Geometry on								
SES(Books)	0.263*	0.238*	0.183*	0.261*	0.257*	0.224*	0.282*	0.251*
Achievement—Number on								
SES(Books)	0.260*	0.234*	0.205*	0.235*	0.274*	0.242*	0.281*	0.256*
Between level (Class level)								
Achievement—Data on								
SES(Books)	0.660*	0.325*	0.205	0.802*	0.605*	0.680*	0.561*	0.846*
Content coverage: Number	0.054	0.077	0.051	0.077	− 0.074	− 0.007	− 0.077	0.072
Content coverage: Geometry	− 0.036	0.213*	0.209	0.052	0.160*	0.000	0.126	0.111
Content coverage: Data	0.163	− 0.018	0.036	− 0.005	− 0.013	0.079	0.306*	0.004
Achievement—Geometry on								
SES(Books)	0.606*	0.349*	0.273*	0.849*	0.565*	0.674*	0.573*	0.801*
Content coverage: Number	0.051	0.100	0.081	0.079	− 0.084	− 0.011	− 0.110	0.095
Content coverage: Geometry	− 0.002	0.242*	0.222	0.016	0.138*	0.021	0.165	0.094

(continued)

(continued)

	2011				2015			
	Denmark	Finland	Norway	Sweden	Denmark	Finland	Norway	Sweden
Content coverage: Data	0.151	0.005	−0.017	−0.028	−0.006	0.047	0.274*	0.048
Achievement—Number on								
SES(Books)	0.649*	0.385*	0.351*	0.779*	0.528*	0.719*	0.536*	0.838*
Content coverage: Number	0.067	0.100	0.047	0.152	−0.098	0.034	−0.086	0.095
Content coverage: Geometry	−0.026	0.217*	0.225*	0.034	0.182*	−0.034	0.153	0.088
Content coverage: Data	0.146	−0.046	−0.019	0.030	−0.011	0.040	0.328*	0.014
Model fit								
$\chi^2 = (df, n) = $ mean	(0, 2859) = 0.054	(0, 4323) = 0.003	(0, 2527) = 0.106	(0, 2961) = 12.254	(0, 3171) = 0.089	(0, 4959) = 0.008	(0, 3854) = 0.319	(0, 3843) = 0.224
RMSEA	0.000	0.000	0.000	0.000	0.000	0.000	0.000	0.000
CFI	1.000	1.000	1.000	0.998	1.000	1.000	1.000	1.000
TLI	1.000	1.000	1.000	1.000	1.000	1.000	1.000	1.000
SRMR (within)	0.000	0.000	0.000	0.000	0.000	0.000	0.000	0.000
SRMR (between)	0.001	0.001	0.001	0.011	0.001	0.001	0.001	0.000
Within level (Student level)								
Achievement—Life Science on								
SES(Books)	0.262*	0.248*	0.285*	0.304*	0.275*	0.242*	0.269*	0.263*
Achievement—Physical Science on								
SES(Books)	0.256*	0.259*	0.224*	0.266*	0.270*	0.230*	0.285*	0.272*
Achievement—Earth Science on								
SES(Books)	0.237*	0.212*	0.292*	0.275*	0.268*	0.265*	0.221*	0.279*

(continued)

(continued)

	2011				2015			
	Denmark	Finland	Norway	Sweden	Denmark	Finland	Norway	Sweden
Between level (Class level)								
Achievement—Life Science on								
SES(Books)	0.911*	0.373*	0.332*	0.870*	0.736*	0.794*	0.543*	0.837*
Content coverage: Life Science	0.357*	0.023	0.118	− 0.010	− 0.121	0.229*	0.122	− 0.034
Content coverage: Physical Science	− 0.072	− 0.021	− 0.105	0.074	− 0.149	− 0.144	0.066	− 0.016
Content coverage: Earth Science	0.181	0.125	0.048	0.083	0.157	− 0.016	0.107	0.149
Achievement—Physical Science on								
SES(Books)	0.902*	0.424*	0.340*	0.884*	0.738*	0.693*	0.604*	0.825*
Content coverage: Life Science	0.326*	0.036	0.084	− 0.011	− 0.096	0.248*	0.169	− 0.080
Content coverage: Physical Science	− 0.072	− 0.033	− 0.117	0.114	− 0.182	− 0.092	0.075	0.009
Content coverage: Earth Science	0.200	0.124	0.049	0.049	0.157	− 0.033	0.055	0.150
Achievement—Earth Science on								
SES(Books)	0.894*	0.383*	0.432*	0.861*	0.700*	0.809*	0.596*	0.826*
Content coverage: Life Science	0.304*	0.006	0.017	0.007	− 0.129	0.186*	0.134	− 0.086
Content coverage: Physical Science	− 0.048	− 0.062	− 0.093	0.058	− 0.176	− 0.102	0.091	− 0.015

(continued)

(continued)

	2011				2015			
	Denmark	Finland	Norway	Sweden	Denmark	Finland	Norway	Sweden
Content coverage: Earth Science	0.189	0.160	0.092	0.062	0.188	0.016	0.073	0.166
Model fit								
$\chi^2 = $ (df, n) = mean	(0, 1428) = 0.252	(0, 4147) = 0.061	(0, 2022) = 0.265	(0, 2575) = 0.181	(0, 1298) = 0.097	(0, 4864) = 0.013	(0, 2541) = 0.146	(0, 3061) = 0.240
RMSEA	0.000	0.000	0.000	0.000	0.000	0.000	0.000	0.000
CFI	1.000	1.000	1.000	1.000	1.000	1.000	1.000	1.000
TLI	1.000	1.000	1.000	1.000	1.000	1.000	1.000	1.000
SRMR (within)	0.000	0.000	0.000	0.000	0.000	0.000	0.000	0.000
SRMR (between)	0.002	0.001	0.005	0.001	0.001	0.001	0.001	0.000

	2019			
	Denmark	Finland	Norway	Sweden
Within level (Student level)				
Achievement—Data on				
SES(Books)	0.204*	0.272*	0.228*	0.256*
Achievement—Geometry on				
SES(Books)	0.262*	0.280*	0.267*	0.278*
Achievement—Number on				
SES(Books)	0.240*	0.273*	0.256*	0.270*
Between level (Class level)				
Achievement—Data on				
SES(Books)	0.529	0.830*	0.635*	0.734*
Content coverage: Number	0.209	0.061	0.226	0.013

(continued)

(continued)

| | 2019 | | | |
	Denmark	Finland	Norway	Sweden
Content coverage: Geometry	0.081	0.062	− 0.092	0.144
Content coverage: Data	0.131	− 0.083	0.167	0.002
Achievement—Geometry on				
SES(Books)	0.524	0.835*	0.715*	0.816*
Content coverage: Number	0.172	0.055	0.171	0.057
Content coverage: Geometry	0.144	0.030	− 0.044	0.185*
Content coverage: Data	0.154	− 0.123	0.084	− 0.028
Achievement—Number on				
SES(Books)	0.571	0.841*	0.602*	0.777*
Content coverage: Number	0.197	0.073	0.264	0.124
Content coverage: Geometry	0.084	0.025	− 0.030	0.161
Content coverage: Data	0.139	− 0.104	0.155	− 0.056
Model fit				
$\chi^2 = (df, n) = mean$	(0, 2582) = 0.086	(0, 4661) = 0.268	(0, 2808) = 0.238	(0, 3469) = 0.062
RMSEA	0.000	0.000	0.000	0.000
CFI	1.000	1.000	1.000	1.000
TLI	1.000	1.000	1.000	1.000
SRMR (within)	0.000	0.000	0.000	0.000
SRMR (between)	0.002	0.001	0.001	0.000

(continued)

(continued)

	2019			
	Denmark	Finland	Norway	Sweden
Within level (Student level)				
Achievement—Life Science on				
SES(Books)	0.305*	0.295*	0.320*	0.293*
Achievement—Physical Science on				
SES(Books)	0.301*	0.254*	0.259*	0.281*
Achievement—Earth Science on				
SES(Books)	0.290*	0.312*	0.315*	0.278*
Between level (Class level)				
Achievement—Life Science on				
SES(Books)	0.734*	0.767*	0.705*	0.878*
Content coverage: Life Science	−0.052	0.056	−0.161	0.007
Content coverage: Physical Science	0.016	0.133	0.181	0.181
Content coverage: Earth Science	0.031	−0.112	0.074	−0.096
Achievement—Physical Science on				
SES(Books)	0.746*	0.756*	0.731*	0.860*
Content coverage: Life Science	0.094	0.095	−0.193	0.005
Content coverage: Physical Science	−0.001	0.138	0.158	0.188*
Content coverage: Earth Science	−0.012	−0.168*	0.088	−0.109
Achievement—Earth Science on				
SES(Books)	0.695*	0.796*	0.758*	0.871*

(continued)

(continued)

| | 2019 | | | |
	Denmark	Finland	Norway	Sweden
Content coverage: Life Science	0.029	0.096	−0.169	0.030
Content coverage: Physical Science	−0.035	0.118	0.192	0.198*
Content coverage: Earth Science	−0.067	−0.161*	0.078	−0.116
Model fit				
$\chi^2 = (df, n) = $ mean	(0, 1569) = 0.124	(0, 4518) = 0.074	(0, 2035) = 0.130	(0, 2633) = 0.151
RMSEA	0.000	0.000	0.000	0.000
CFI	1.000	1.000	1.000	1.000
TLI	1.000	1.000	1.000	1.000
SRMR (within)	0.000	0.000	0.000	0.000
SRMR (between)	0.002	0.001	0.001	0.001

NB *P < 0.05

Appendix 4 Bar Graphs Showing Achievement in Science Domains for the Years 2011 to 2019 for All Countries

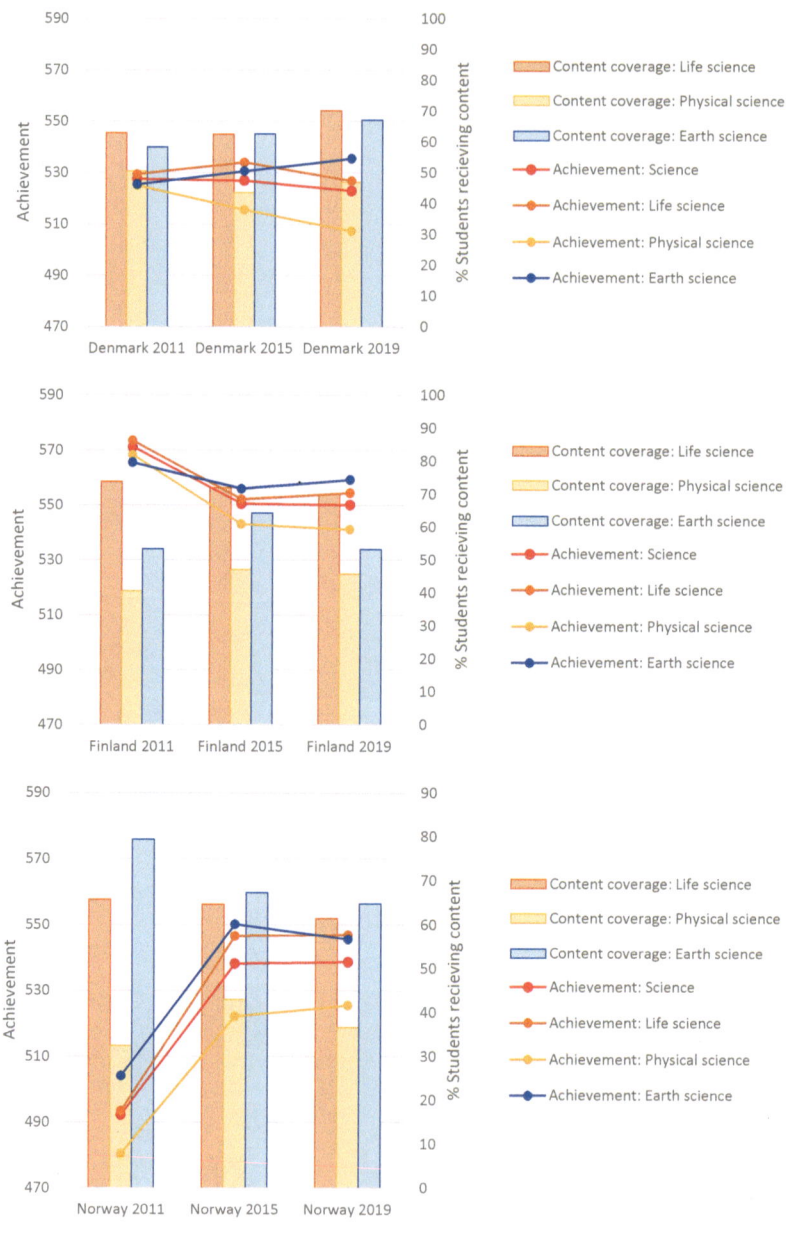

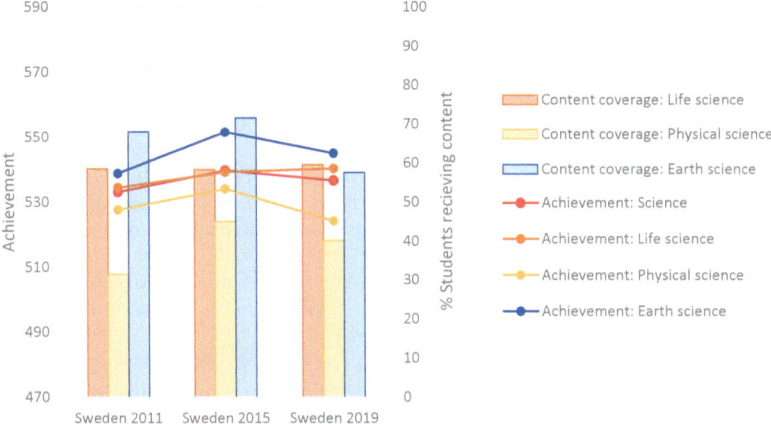

Note One bar graph per country.

References

Martin, M. O., Mullis, I. V. S., Foy, P., & Hooper, M. (2016). *TIMSS 2015 international results in science.* TIMSS & PIRLS International Study Center, Boston College. https://timssandpirls.bc.edu/timss2015

Martin, M.O., Mullis, I.V.S., Foy, P., & Stanco, G.M. (2012). *TIMSS 2011 international results in science.* TIMSS & PIRLS International Study Center, Boston College. https://timssandpirls.bc.edu/timss2011/international-results-science.html

Mullis, I.V.S., Martin, M.O., Foy, P., & Arora, A. (2012). *TIMSS 2011 international results in mathematics.* TIMSS & PIRLS International Study Center, Boston College. https://timssandpirls.bc.edu/timss2011/international-results-mathematics.html

Mullis, I. V. S., Martin, M. O., Foy, P., & Hooper, M. (2016). *TIMSS 2015 international results in mathematics.* TIMSS & PIRLS International Study Center, Boston College. https://timssandpirls.bc.edu/timss2015/international-results/timss-2015/mathematics/student-achievement/

Mullis, I. V. S., Martin, M. O., Foy, P., Kelly, D. L., & Fishbein, B. (2020). *TIMSS 2019 international results in mathematics and science.* TIMSS & PIRLS International Study Center, Boston College. https://timssandpirls.bc.edu/timss2019/international-results/

References

Atlay, C., Tieben, N., Hillmert, S., & Fauth, B. (2019). Instructional quality and achievement inequality: How effective is teaching in closing the social achievement gap? *Learning and Instruction, 63*, 101211. https://doi.org/10.1016/j.learninstruc.2019.05.008

Daus, S., Nilsen, T., & Braeken, J. (2019). Exploring content knowledge: Country profile of science strengths and weaknesses in TIMSS. Possible implications for educational professionals and science research. *Scandinavian Journal of Educational Research*, *63*(7), 1102–1120. https://doi. org/10.1080/00313831.2018.1478882

Icelandic Ministry of Education Science and Culture. (2014). *The Icelandic national curriculum guide for compulsory schools—with subjects areas*. Ministry of Education, Science and Culture. https://www.government.is/library/01-Ministries/Ministry-of-Education/Curriculum/ adalnrsk_greinask_ens_2014.pdf

Kelly, D. L., Centurino, V. A., Martin, M. O., & Mullis, I. V. (Eds.). (2020). *TIMSS 2019 Encyclopedia: Education policy and curriculum in mathematics and science*. TIMSS & PIRLS International Study Center, Boston College. https://timssandpirls.bc.edu/timss2019/encyclopedia/

Kjeldsen, C. C., Kristensen, R. M., & Christensen, A. A. (2020). *Matematik og natur/teknologi i 4. klasse: Resultater af TIMSS-undersøgelsen 2019*. Aarhus Universitetsforlag. https://unipress. dk/udgivelser/m/matematik-og-naturteknologi-i-4-klasse/

Martin, M. O., Mullis, I. V., Foy, P., & Hooper, M. (2016). *TIMSS 2015 international results in science*. TIMSS & PIRLS International Study Center, Boston College. http://timssandpirls.bc. edu/timss2015/international-results/

Martin, M. O., Mullis, I. V., Foy, P., & Stanco, G. M. (2012). *TIMSS 2011 international results in science*. TIMSS & PIRLS International Study Center, Boston College. https://timssandpirls.bc. edu/timss2011/international-results-science.html

Martin, M. O., von Davier, M., & Mullis, I. V. S. (Eds.). (2020). *Methods and procedures: TIMSS 2019 technical report*. TIMSS & PIRLS International Study Center, Boston College. https://tim ssandpirls.bc.edu/timss2019/methods

Mullis, I. V., & Martin, M. O. (Eds.). (2013). *TIMSS 2015 assessment frameworks*. TIMSS & PIRLS International Study Center, Boston College. https://timssandpirls.bc.edu/timss2015/fra meworks.html

Mullis, I. V., & Martin, M. O. (Eds.). (2017). *TIMSS 2019 assessment frameworks*. TIMSS & PIRLS International Study Center, Boston College. https://timssandpirls.bc.edu/timss2019/fra meworks/

Mullis, I. V., Martin, M. O., Foy, P., & Arora, A. (2012a). *TIMSS 2011 international results in mathematics*. TIMSS & PIRLS International Study Center, Boston College. https://timssandp irls.bc.edu/timss2011/international-results-mathematics.html

Mullis, I. V., Martin, M. O., Foy, P., & Hooper, M. (2016a). *TIMSS 2015 international results in mathematics*. TIMSS & PIRLS International Study Center, Boston College. http://timssandp irls.bc.edu/timss2015/international-results/

Mullis, I. V., Martin, M. O., Foy, P., Kelly, D. L., & Fishbein, B. (2020). *TIMSS 2019 international results in mathematics and science*. TIMSS & PIRLS International Study Center, Boston College. https://timssandpirls.bc.edu/timss2019/international-results

Mullis, I. V., Martin, M. O., Goh, S., & Cotter, K. (Eds.). (2016b). *TIMSS 2015 encyclopedia: Education policy and curriculum in mathematics and science*. TIMSS & PIRLS International Study Center, Boston College. http://timssandpirls.bc.edu/timss2015/encyclopedia/

Mullis, I. V., Martin, M. O., Minnich, C. A., Stanco, G. M., Arora, A., Centurino, V. A., & Castle, C. E. (Eds.). (2012b). *TIMSS 2011 encyclopedia: Education policy and curriculum in mathematics and science. Volume 1 & 2*. TIMSS & PIRLS International Study Center, Boston College. https:// timssandpirls.bc.edu/timss2011/encyclopedia-timss.html

Mullis, I. V., Martin, M. O., Ruddock, G. J., O'Sullivan, C., & Preuschoff, C. (Eds.). (2009). *TIMSS 2011 assessment frameworks*. International Association for the Evaluation of Educational Achievement. https://timssandpirls.bc.edu/timss2011/downloads/TIMSS2011_Framew orks.pdf

Scheerens, J. (2017). Conceptualization. In J. Scheerens (Ed.), *Opportunity to learn, curriculum alignment and test preparation : A research review* (pp. 7–22). Springer International Publishing. https://doi.org/10.1007/978-3-319-43110-9_2

Wagner, J. -P., & Hastedt, D. (2022). Valuing curriculum-based international large-scale assessments: Ensuring alignment with national curricula in IEA studies. *IEA Compass: Briefs in*

Education. Number 16. International Association for the Evaluation of Educational Achievement. https://www.iea.nl/index.php/publications/series-journals/iea-compass-briefs-education-series/march-2022-valuing-curriculum

Rune Müller Kristensen (born in 1975) is an associate professor of sociology of education at the Danish School of Education, Aarhus University. He has been actively engaged in research as part of the Danish contribution to the TIMSS project since 2018. His primary research interests lie in the domains of International Large-Scale Assessments, as well as research on teachers and teacher education.

Victoria Rolfe (born 1987) is a senior lecturer at the Department of Education and Special Education, University of Gothenburg. Her research focus is on educational equity, educational measurement, and the opportunity to learn.

Chapter 5
Teaching Quality and Assessment Practice: Trends Over Time and Correlation with Achievement

Nani Teig⬤ and Jennifer Maria Luoto⬤

5.1 Introduction

Examining teaching quality and assessment practices in primary schools is of paramount importance, particularly in the context of mathematics and science education. These subjects play a significant role in fostering the development of problem-solving and critical thinking skills, which are crucial for students' academic and long-term success (Delahunty et al., 2020). Although there has been a growing interest in examining teaching quality and assessment practice as key factors influencing student learning outcomes (Andrade, 2019; Klieme & Nilsen, 2022), few studies have compared these constructs in primary mathematics and science classrooms across Nordic countries. This chapter aims to contribute to this expanding field of research by investigating the trends in teaching quality and assessment practice over time as well as their relations to student achievement in mathematics and science across Nordic countries (i.e., Denmark, Finland, Norway, and Sweden). The findings can offer valuable insights for policymakers and practitioners in designing and implementing evidence-based policies and interventions that promote high-quality teaching and assessment practices. Ultimately, this research seeks to support the continuous improvement of education systems in Nordic countries, enabling students to reach their full potential through mathematics and science classrooms.

N. Teig (✉) · J. M. Luoto
Department of Teacher Education and School Research, University of Oslo, Postboks 1099, Blindern, Oslo, Norway
e-mail: nani.teig@ils.uio.no

J. M. Luoto
e-mail: j.m.luoto@ils.uio.no

N. Teig et al. (eds.), *Effective and Equitable Teacher Practice in Mathematics and Science Education*, IEA Research for Education 14,
https://doi.org/10.1007/978-3-031-49580-9_5

155

5.2   The Nordic Educational Contexts: Mathematics and Science Education

Nordic countries share similar historical, cultural, and economic characteristics, suggesting that schools and teachers operate under relatively comparable conditions (Teig & Steinmann, 2023). Despite these similarities, significant differences exist in the trends of average mathematics and science achievement across Nordic countries (see Chap. 1). While trends in average achievement have largely remained stable in Norway and Denmark, these trends have been decreasing in Finland and increasing in Sweden over several of IEA's Trends in International Mathematics and Science Study (TIMSS) cycles (Mullis et al., 2020).

Previous research has highlighted performance level differences across schools and classrooms within Nordic countries (Yang Hansen et al., 2014). In Finland, no school-level differences existed in grades four and eight, while Norway and Sweden displayed substantial school-level differences for both grades (Yang Hansen et al., 2014). Classroom-level differences also existed in these countries, with Finland showing considerable variation in performance compared to Norway and Sweden (Yang Hansen et al., 2014). While classroom differences may arise from the sorting of students into various classes, they could also reflect the disparities in classroom activities and student experiences (Creemers & Kyriakides, 2008). Teachers play a crucial role in providing learning opportunities through instruction and assessment, allowing them to monitor and improve student performance. Given the differences in achievement levels and the potential role of teaching quality and assessment practices in shaping student performance, it is increasingly important to examine these factors over time in Nordic countries.

5.3   Teaching Quality and Student Achievement

Teaching quality is a multidimensional construct and generally considered to be teaching practices that are related to some types of students' cognitive and non-cognitive outcomes (Baumert et al., 2010; Klieme et al., 2009; see Chap. 2 Theoretical Framework of Teacher Practice for further details). The conceptualization of teaching quality in this book is closely aligned with the Three Basic Dimensions (TBD) of teaching quality: classroom management, supportive climate, and cognitive activation (Klieme et al., 2009).

Classroom management refers to the strategies, techniques, and processes that teachers use to create and maintain a well-organized, focused, and orderly learning environment with minimal disruptions (Praetorius et al., 2018). This includes managing instructional time effectively, ensuring that students stay on task, implementing clear rules, and maintaining order and discipline (Marder et al., 2023).

Supportive climate pertains to the quality of interactions in the classroom and encompasses various aspects, including teacher support, classroom interaction (teacher–student and student–student relationships), and instructional clarity (Nilsen et al., 2016; Praetorius et al., 2018). Supportive climate includes addressing individual student needs, offering various learning opportunities and engaging materials, helping students to understand and link new concepts, clarifying conceptual misunderstandings, and setting clear expectations (Nilsen et al., 2016). This chapter focuses specifically on teacher support and instructional clarity.

Cognitive activation involves instructional approaches and learning tasks that stimulate students' cognitive processing, promote conceptual understanding, and encourage students to engage in higher-order thinking (Baumert et al., 2010; Förtsch et al., 2017; Klieme et al., 2009). The level of cognitive activation often depends on task selection and implementation in the classrooms (Baumert et al., 2010; Lipowsky et al., 2009).

Previous studies investigating the correlation between teaching quality and student achievement in mathematics and science using survey data have produced mixed results. Reviewing studies that investigate these correlations with TIMSS and the OECD's Programme for International Student Assessment (PISA) data, Klieme and Nilsen (2022), highlighted that most studies demonstrated a small positive correlation between teaching quality and achievement. (e.g., Bellens et al., 2019). For classroom management, findings have been inconsistent when correlating with student achievement (e.g., Bellens et al., 2019). Supportive climate displayed the strongest correlation with achievement among the TBD dimensions, while cognitive activation showed a positive correlation with mathematics achievement but often had a negative or insignificant correlation with science achievement when considering inquiry practices. In another review by Klieme (2019), classroom management was positively related to achievement in both subjects across all countries, after controlling for student background and school composition. In contrast, teacher support was not significantly related or only spuriously related to achievement. Cognitive activation showed a slightly positive correlation with mathematics achievement in most countries. However, inquiry-based teaching, as a form of cognitive activation, was negatively associated with science achievement. Both review studies also showed that the relations between TBD dimensions and student outcomes may vary across the teaching activities representing the TBD dimensions, subjects, and countries (Klieme & Nilsen, 2022). Hence, studies examining these relationships should account for and clearly explain these variations.

Focusing on Nordic countries (excluding Iceland), Nilsen et al. (2018) examined teachers' perceptions of teaching quality related to cognitive activation and teacher support in science for fourth and eighth grades using TIMSS 2015 data. While teaching quality was found to have a positive correlation with student achievement, the strength of the correlation, however, varied across the Nordic countries. In Denmark, science teachers' self-reported instructional quality showed no significant relationship with student achievement, while in Finland, Sweden and Norway, teaching quality was positively and significantly correlated with fourth-grade science achievement. For grade eight,[1] Nilsen et al. (2018) revealed a significant correlation between teaching quality and student achievement in science in Norway and Sweden. Furthermore, Teig and Nilsen (2022) investigated the profiles of science teaching quality focusing on teacher support and instructional clarity by exploring Norwegian students' perceptions of teaching quality in grades five and nine using TIMSS 2015 data. They found that teaching quality patterns varied across both grades. In general, students who perceived their teachers as having high teaching quality were somewhat more likely to have higher science achievement.

These studies also highlight the importance to exercise caution when interpreting the relationships between teaching quality and learning outcomes due to the potential for reverse causality (e.g., Nilsen et al., 2018; Teig & Nilsen, 2022). For instance, students with low achievement and negative attitudes towards schooling might perceive their teachers' instruction as low quality or more cognitively challenging compared to others. This complexity highlights the need to further explore the intricate relationship between teaching quality and student outcomes.

5.4 Teacher Assessment Practice and Student Achievement

Teacher assessment practice encompasses various methods and strategies that teachers use to gather evidence of students' current understanding and use it to inform educational decisions, such as in planning lessons, adapting instruction, selecting assignments, providing feedback, and assigning grades (Black & Wiliam, 2009; Gardner et al., 2010; Herppich et al., 2018). These assessments can take multiple forms, including formal assessments such as tests and quizzes, as well as informal assessments like observations and discussions with students.

The primary goal of classroom-level assessments is to provide support to both teachers and students as they work towards determining, monitoring, and enhancing performance (Andrade & Brookhart, 2020; Gardner et al., 2010). By employing effective assessment practices, teachers can better identify their students' strengths and weaknesses, gain valuable insights for adapting their instructional approaches, and empower students with a clearer understanding of the steps they need to take in order to improve their learning (Andrade & Brookhart, 2020; Kanjee, 2009). A holistic assessment practice facilitates a positive collaboration between teachers and students, leading to the development of engaging and effective learning environments.

Research has demonstrated that teacher assessment practice is an essential component of effective teaching and learning, as it often significantly impacts student outcomes (Andersson & Palm, 2017; Hattie, 2009; Palm et al., 2017; Panadero et al., 2017). Palm et al. (2017) conducted a review that revealed a positive relationship between student achievement in mathematics and three types of teacher assessment practice: feedback, student self-assessment, and teacher assessment with subsequent

---

[1] Only Norway and Sweden participated grade eight survey in TIMSS 2015.

instructional actions. In contrast, Mostafa et al. (2018) discovered a negative relationship between teacher feedback and science performance in nearly all countries participating in PISA 2015. When it comes to students' perceptions of teacher feedback in Nordic countries, students in Denmark, Finland, and Iceland seemed to perceive feedback less frequently than those in Norway and Sweden (Sortkær, 2019). Additionally, Nordic students perceived less feedback than students from other OECD countries (Sortkær, 2019). A recent study also showed that a high frequency of teacher feedback in Nordic countries was more commonly reported in low-achieving students and schools (Rohatgi et al., 2022).

Homework is a common practice that holds significant potential as an assessment tool when used for monitoring student learning in mathematics and science (Martin et al., 2016). While it may not be immediately apparent, homework can indeed be considered an assessment practice. This is because homework assignments allow teachers to evaluate a student's understanding and application of the content covered in the classroom. Moreover, it provides an opportunity for students to assess their own learning progress, identify gaps in their understanding, and practice problem-solving skills (Fan et al., 2017; Fernández-Alonso & Muñiz, 2022). Despite its widespread use, homework showed limited effects on student learning (Fan et al., 2017; Scheerens, 2016). This could be attributed to whether or not homework is used strategically to assess students' developing knowledge, such as identifying the specific types of tasks that may pose challenges.

The use of homework in Nordic classrooms has not been extensively studied, yet it is often a topic of political debate. In Finland, homework in mathematics is an integrated classroom practice at the lower secondary level, with lessons typically starting with homework review and ending with new homework assignments (Krzywacki et al., 2016; Luoto et al., 2022). This pattern suggests a cultural tradition of homework routines (Fernández-Alonso & Muñiz, 2020). Investigating homework as a potentially important part of teacher assessment practice can provide valuable insights into its cultural tradition and contribute to the ongoing debate on the efficacy of homework in promoting student learning.

Previous research highlights the complex relationship between assessment practices and student achievement, emphasizing the importance of context and subject matter. There is a clear need for conducting research that examines teacher assessment practice over time, especially in Nordic countries, to better understand the nuances of these relationships and to develop optimal assessment strategies for various subjects and educational contexts. Investigating the relations between teacher assessment practices and student achievement is also crucial, as it can provide invaluable insights into the effectiveness of current practices and help identify areas for improvement.

5.5   The Present Study

In the present study, we utilize data from TIMSS 2011, 2015, and 2019 in mathematics and science for grade four across Nordic countries to address the following research questions (RQs):

RQ 1. What are the trends in teaching quality and assessment practice over time?
RQ 2. What is the relationship between teaching quality and student achievement?
RQ 3. What is the relationship between teacher assessment practice and student achievement?

5.6 Methods

5.6.1 Data and Variables

Data for the analyses were drawn from TIMSS, a large-scale international survey that assesses student performance in mathematics and science at the fourth- and eighth-grade levels in participating countries every fourth year. To examine trends in teaching quality and assessment practices over time (RQ 1), data from TIMSS 2011 to 2019 were analyzed. Additionally, the study focused on the TIMSS 2019 grade four data to investigate the relations between teaching quality and achievement (RQ 2) and the relations between teacher assessment practice and student achievements (RQ 3).

Table 5.1 summarizes the different aspects of teaching quality and assessment practice that were addressed in the specific RQs and TIMSS data. It is important to note that only a few items measuring teaching quality and assessment practice were consistent across TIMSS 2011 to 2019. As a result, RQ 1 had less comprehensive coverage of these constructs compared to RQs 2 and 3, which examined the most recent TIMSS cycle in 2019. Further details on the specific items used to measure the constructs are presented in Sect. 5.7 Findings and Appendix 1.

Teaching quality

TIMSS assessed teaching quality using both student and teacher background questionnaires, examining various aspects of classroom management, teacher support and instructional clarity, and cognitive activation. To assess *classroom management*, the student questionnaire measured the frequency of various disruptive and disorderly behaviors in the mathematics classroom using six items (e.g., "my teacher has to keep telling us to follow the classroom rules" or "students interrupt the teacher") with a response scale: every or almost every lesson, about half the lessons, some lessons, and never. *Teacher support and instructional clarity*, as important aspects of supportive climate, were measured using student agreement on various statements (e.g., "my teacher does a variety of things to help us learn" or "my teacher is easy to understand") with a response scale that ranges from agree a lot to disagree a lot.

The manifestation of *cognitive activation* varies depending on the subject. Since a distinction can be made between general and subject-specific cognitive activation (Teig et al., 2019), this study operationalizes cognitive activation differently for mathematics and science. In mathematics, teachers were asked on how often they engaged students in generic cognitive activation (e.g., "relate the lesson to students' daily lives") or problem-solving (e.g., "apply what students have learned to new

Table 5.1 The aspects of teaching quality and assessment practice across RQs and TIMSS data

Construct	Questionnaire		RQ 1 Trends in teaching quality and assessment practice (TIMSS 2011–2019)	RQ 2 Relations between teaching quality and student achievement (TIMSS 2019)	RQ 3 Relations between assessment practice and student achievement (TIMSS 2019)
	Student	Teacher			
Teaching quality					
Classroom management	✓			✓	
Teacher support and clarity of instruction	✓		✓	✓	
Cognitive activation	✓	✓	✓	✓	
Assessment practice					
Homework frequency		✓	✓		✓
Homework time		✓	✓		✓
In-class homework discussion		✓	✓		✓
Teacher emphasis on assessment strategies		✓			✓

problem situations on their own"). In science, teachers were asked how often they engaged students in various inquiry-based cognitive activation (e.g., "design or plan experiments or investigations"). The items related to cognitive activation employ a frequency-based response scale, ranging from "every or almost every lesson" to "never".

To supplement the teacher questionnaire, this study also used students' responses on the frequency with which they conduct experiments in their science lessons to represent inquiry-based cognitive activation. The response scale for this component ranged from "never", "a few times a year", "once or twice a month", to "at least once a week". This approach provides a more comprehensive understanding of the role of cognitive activation in different subject areas.

Teacher assessment practice

As shown in Table 5.1, only homework frequency, the time needed to complete homework, and in-class homework discussion were measured repeatedly across TIMSS 2011–2019. New items representing how much importance teachers place on various assessment strategies in mathematics and science were first introduced in TIMSS 2019 (e.g., "asking students to answer questions during class"). Further details on the specific items used are presented in Sect. 5.7 Findings and Appendix 2.

Student achievement in mathematics and science

TIMSS assessed student achievement with a standardized test that covers cognitive domains and subject-specific content domains. Student achievement was estimated via a measurement model that produced a set of five plausible values to represent the likely distribution of student performance. All plausible values were incorporated into the analyses to produce an average of the model estimates and adjusted standard errors (see Chap. 3 Analytical Framework).

5.6.2 Data Analysis

Data were first prepared using the IDB Analyzer 4.0, while the main analyses were conducted using Mplus 8.8 (Muthén & Muthén, 2022) and IBM SPSS Statistics 28.0. To address RQ 1 about the trends in teaching quality and assessment practice over time, descriptive statistics and one-way ANOVAs were performed to test for significant mean differences within-country across three pairwise comparisons: TIMSS 2011 versus 2015, 2015 versus 2019, and 2011 versus 2019.

To investigate the relationship between teaching quality and student achievement (RQ 2) and the relationship between teacher assessment practice and student achievement (RQ 3), we implemented a two-level approach to account for TIMSS' cluster sampling design in which students were nested within classrooms. Specifically, we employed multilevel structural equation modelling (MSEM) with students nested in classrooms. The MSEM approach served two main purposes: (a) established measurement models to represent teaching quality dimensions; (b) examined the relations between teaching quality, assessment practice, and student achievement. To accomplish (a), we used multilevel confirmatory factor analysis (MCFA)—an extension of CFA, to multilevel situations that have proved useful to study the factor structure of instructional practices at two levels (Brown, 2015; Morin et al., 2014). Model fit was evaluated using Ryu's (2014) partial saturation approach by obtaining test statistics and fit indices for each level separately. We referred to common guidelines for an acceptable model fit (i.e., CFI ≥ 0.95, TLI ≥ 0.95, RMSEA ≤ 0.08, and SRMR ≤ 0.10; Marsh et al., 2005).

We further performed MSEM with a multi-group approach to separately examine the distinct relationships between teaching quality, assessment practice, and student achievement in each of the Nordic countries. Measurement invariance was conducted to ensure valid comparisons across groups (Sass & Schmitt, 2013). The data generally supported sufficient levels of measurement invariance (further details are presented in Chap. 3 Analytical Framework).

Analyses for RQ 2 were conducted at both the student and classroom levels, with two exceptions: (a) the relationship between the construct cognitive activation and student achievement, and (b) the relationship between assessment practice and student achievement, both of which were analyzed only at the classroom level.

5.7 Findings

5.7.1 The Trends in Teaching Quality and Assessment Practice Over Time

Figures 5.1, 5.2, 5.3, and 5.4 provide a graphical representation of the mean differences of these items across two-time TIMSS cycles (i.e., TIMSS 2011 versus 2015, 2015 versus 2019, and 2011 versus 2019).

Teaching quality

Classroom management was first introduced in TIMSS 2019 and is not included in the analyses of the mean differences between 2011 and 2019.

Two items were related to supportive climate and directly related to teacher support and instructional clarity in mathematics and science. As shown in Fig. 5.1, Nordic countries experienced an overall decline in teacher support and instructional clarity from 2011 to 2019. This decrease was more pronounced between 2015 and 2019 than in 2011 to 2015 in Denmark and Norway, while in Sweden, a different pattern was observed. Finland exhibited a significant increase between 2011 and 2015, followed by a decrease from 2015 to 2019. Similar patterns were identified in both mathematics and science.

Cognitive activation was measured using two items for general cognitive activation, three items for mathematics, and five items for science. The mean differences showed mixed patterns in cognitive activation across cycles, countries, and activities in both mathematics and science (Fig. 5.2). For instance, between 2011 and 2015, students' opportunities to observe natural phenomena and describe their observations decreased in Denmark, increased in Finland and Norway, and remained unchanged in Sweden. Between 2015 and 2019, the same activity decreased in Finland, increased in Norway and Sweden, and showed no significant changes in Denmark.

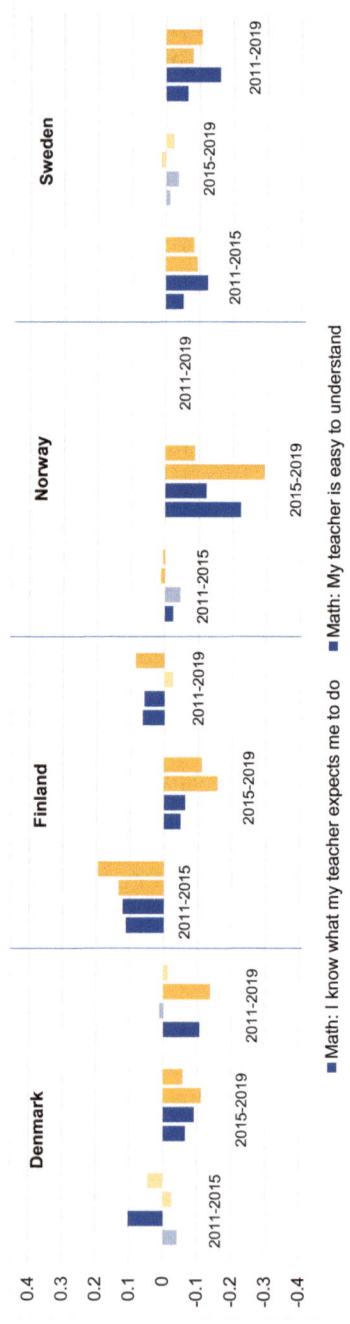

Fig. 5.1 The mean differences in students' perceived teacher support and instructional clarity. *Note* Dark bars = significant difference between cycles (p-value < .05), light bars = non-significant difference between cycles (p-value < .05), positive value = higher mean in the more recent TIMSS cycle of the comparison, negative value = lower mean in the more recent TIMSS cycle of the comparison

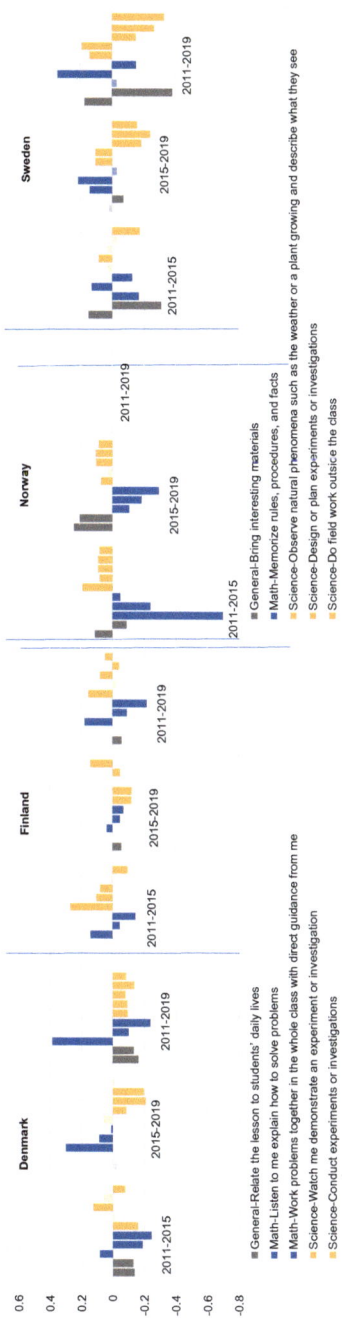

Fig. 5.2 The mean differences in teachers' perceived cognitive activation. *Note* Dark bars = significant difference between cycles (p-value < .05), light bars = non-significant difference between cycles (p-value < .05), positive value = higher mean in the more recent TIMSS cycle of the comparison, negative value = lower mean in the more recent TIMSS cycle of the comparison

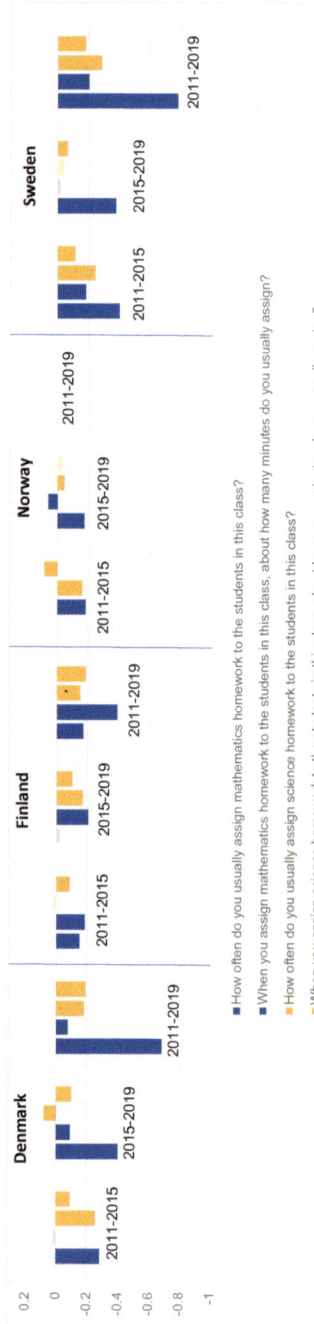

Fig. 5.3 The mean differences in teachers' self-reported assessment practice (homework). *Note* Dark bars = significant difference between cycles (p-value < .05), light bars = non-significant difference between cycles (p-value < .05), positive value = higher mean in the more recent TIMSS cycle of the comparison, negative value = lower mean in the more recent TIMSS cycle of the comparison

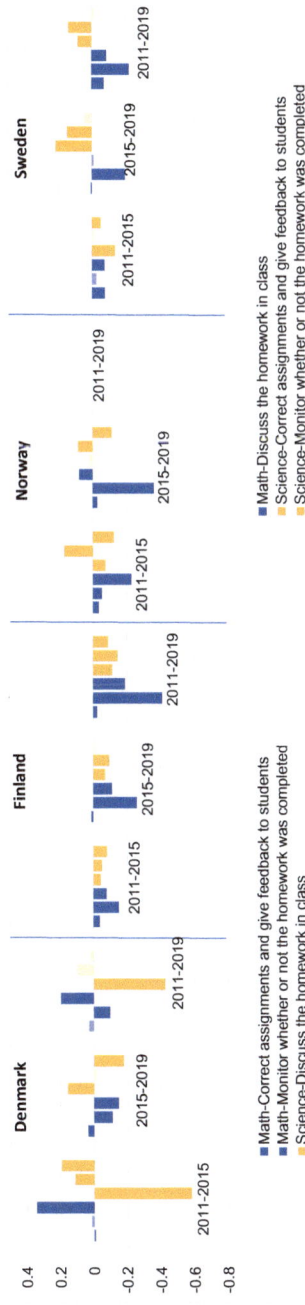

Fig. 5.4 The mean differences in teachers' self-reported assessment practice (the frequency of in-class homework discussion. *Note* Dark bars = significant difference between cycles (p-value < .05), light bars = non-significant difference between cycles (p-value < .05), positive value = higher mean in the more recent TIMSS cycle of the comparison, negative value = lower mean in the more recent TIMSS cycle of the comparison

Assessment practice

Two aspects of teachers' self-reported assessment practice were consistent across the cycles: homework and in-class homework discussion. Figure 5.3 shows an overall decrease in homework frequency and time to complete homework for both subjects across Nordic countries. Another aspect of assessment practice concerns the integration of homework into classroom instruction, represented by three items. Mixed findings were observed for the frequency of in-class homework discussions in the Nordic countries (Fig. 5.4). In Finland, there was a clear decrease in in-class homework discussions from 2011 to 2019. In Sweden, a similar decline occurred in both subjects between 2011 and 2015; however, while in-class homework discussion in mathematics (i.e., monitor whether or not the homework was completed) decreased between 2015 and 2019, the same activity increased in science. The changes varied across cycles, subjects, and even different activities within the same subjects in Denmark and Norway.

5.7.2 The Relationship Between Teaching Quality and Student Achievement

Findings from TIMSS 2019 revealed similar response patterns of teaching quality across Nordic countries (see Appendix 1 for further details on the items used to measure teaching quality). With respect to assessment practice, some variations were found in homework frequency, time to complete homework, and in-class homework discussion, but a similar response pattern was observed in terms of how much teachers place emphasis on various assessment strategies in mathematics and science. As shown in Appendix 2, among Nordic countries, the highest proportion of students taught by teachers who assigned homework was found in Finland. Nevertheless, the majority of Finnish students received homework that took only 15 min or less to complete. Homework assignments were more prevalent in mathematics than in science classrooms across these countries. Additionally, longer tests were more commonly used as assessment strategies in mathematics rather than science lessons, while the opposite was true for long-term projects.

The relations between teaching quality and student achievement

Findings from the MSEM analyses revealed stronger relationships between teaching quality and student achievement in mathematics than science across Nordic countries (Table 5.2). In mathematics, students' perceptions of classroom management were the most robust predictor of achievement compared to other dimensions of teaching quality. Note that classroom management was measured for the first time in TIMSS 2019 and only in mathematics. Perceived teacher support and instructional clarity were related to student achievement in mathematics, especially at the student level. These relations were only significant at the student level for science achievement in Denmark and Finland.

Table 5.2 The relations between teaching quality and student achievement at the student classroom levels

Predictors	Denmark		Finland		Norway		Sweden	
	Student	Class	Student	Class	Student	Class	Student	Class
	β (SE)	β (SE)	β (SE)	β (SE)	β (SE)	β (SE)	β (SE)	β (SE)
Mathematics								
Classroom management	0.07** (0.03)	0.31** (0.11)	0.08** (0.02)	0.14 (0.12)	0.11** (0.03)	0.34** (0.11)	0.08** (0.03)	0.37** (0.09)
Teacher support and instructional clarity	0.22** (0.03)	0.13 (0.13)	0.09** (0.02)	0.20* (0.10)	0.10** (0.02)	0.15 (0.16)	0.06* (0.03)	0.20* (0.10)
Cognitive activation	–	0.22 (0.12)	–	0.14 (0.10)	–	0.02 (0.14)	–	0.12 (0.16)
Science								
Teacher support and instructional clarity	0.09** (0.02)	0.09 (0.12)	0.08** (0.02)	0.09 (0.10)	– 0.02 (0.53)	0.02 (0.91)	– 0.01 (0.03)	0.22 (0.12)
Cognitive activation	–	0.20 (0.18)	–	0.08 (0.11)	–	– 0.09 (0.13)	–	0.10 (0.12)

Note * $p < 0.05$, ** $p < 0.01$

Teachers' perception of cognitive activation was not related to student achievement in both subjects (Table 5.2). However, the findings revealed that the frequency of conducting experiments, measured using the student questionnaire, was related to science achievement in non-linear (inverted U-shape) rather than linear patterns. These relations were observed at the student level in all Nordic countries and at the classroom levels in Finland, Norway, and Sweden.

5.7.3 The Relationship Between Assessment Practice and Student Achievement

As demonstrated in Table 5.3, there were limited associations between teacher assessment practices and student achievement. Homework frequency was positively related to mathematics achievement in Denmark and Sweden, while the time needed to complete homework had a negative relationship with mathematics achievement in Finland. Correcting assignments and providing feedback to students was negatively related to mathematics and science achievement in Finland and Sweden. Notably, no correlations were found between the amount of emphasis teachers placed on various assessment strategies and student achievement in either mathematics or science.

5.8 Discussion and Conclusion

This study examines teaching quality and assessment practices in primary mathematics and science classrooms across Nordic countries (Denmark, Finland, Norway, and Sweden) using TIMSS 2011 to 2019 data. It investigates the trends in teaching quality and assessment practices over time, as well as the relationships of these aspects of teacher practice with student achievement.

5.8.1 The Trends in Teaching Quality and Assessment Practice Over Time

Analyses of TIMSS data from 2011 to 2019 have revealed a decline in aspects of teaching quality related to teacher support and instructional clarity across Nordic countries. One possible explanation for this decline is the changing characteristics of student populations. The TIMSS data showed an increase in the percentages of low socioeconomic status (SES) students and limitations to teaching (Mullis et al., 2020). The percentages of low SES students, as indicated by those who responded to having none or only 1 to 10 books at home, have increased from 7.3 percent in 2011 to 10.2 percent in 2019 (see Chap. 1). Teachers have reported increased limitations

Table 5.3 The relations between assessment practice and student achievement at the classroom level

Predictors	Denmark	Finland	Norway	Sweden
	β (SE)	β (SE)	β (SE)	β (SE)
Mathematics				
Homework frequency	0.23[*] (0.11)	0.74 (0.09)	0.32[*] (0.14)	0.09 (0.11)
Homework time	0.07 (0.11)	− 0.24[*] (0.08)	0.03 (0.13)	− 0.02 (0.08)
In-class homework discussion				
• Correct assignments and give feedback to students	0.07 (0.04)	− 0.09[*] (0.04)	− 0.07 (0.04)	− 0.07 (0.06)
• Discuss the homework in class	0.02 (0.05)	0.04 (0.04)	− 0.03 (0.04)	− 0.15 (0.06)
• Monitor whether or not the homework was completed	0.07 (0.05)	0.06 (0.11)	0.09 (0.08)	− 0.02 (0.07)
Emphasis on assessment strategies				
• Observing students as they work	− 0.02 (0.08)	− 0.11 (0.06)	− 0.02 (0.08)	− 0.10 (0.07)
• Asking students to answer questions during class	− 0.04 (0.05)	0.02 (0.05)	0.00 (0.05)	− 0.07 (0.06)
• Short, regular written assessments	0.02 (0.04)	− 0.03 (0.05)	0.07 (0.07)	0.02 (0.05)
• Longer tests (e.g., unit tests or exams)	0.05 (0.04)	0.09 (0.05)	0.05 (0.04)	0.09 (0.06)
• Long-term projects	0.05 (0.04)	− 0.01 (0.04)	− 0.05 (0.08)	− 0.03 (0.05)
Science				
Homework frequency	0.05 (0.14)	0.11 (0.09)	0.00 (0.18)	0.00 (0.16)
Homework time	0.29 (0.26)	− 0.12 (0.09)	− 0.08 (0.22)	− 0.12 (0.26)
In-class homework discussion				
• Correct assignments and give feedback to students	− 0.04 (0.11)	− 0.08[*] (0.03)	− 0.04 (0.07)	− 0.20[*] (0.07)
• Discuss the homework in class	0.02 (0.21)	0.06 (0.06)	− 0.05(0.07)	0.20 (0.30)
• Monitor whether or not the homework was completed	0.01 (0.21)	− 0.02 (0.07)	0.04 (0.07)	− 0.18 (0.10)
Emphasis on assessment strategies				
• Observing students as they work	− 0.03 (0.07)	− 0.16 (0.05)	− 0.00 (0.09)	− 0.02 (0.09)
• Asking students to answer questions during class	0.03 (0.07)	− 0.09 (0.06)	0.04 (0.11)	− 0.14 (0.09)
• Short, regular written assessments	− 0.05 (0.07)	− 0.03 (0.04)	− 0.04 (0.07)	− 0.07 (0.07)
• Longer tests (e.g., unit tests or exams)	0.04 (0.07)	0.08 (0.04)	0.02 (0.08)	0.00 (0.07)
• Long-term projects	0.04 (0.06)	0.03 (0.04)	− 0.07 (0.08)	− 0.01 (0.05)

[*] $p < 0.05$,

to teaching, including students lacking prior knowledge, being tired, hungry, and causing more disturbances (see Chap. 7). The increasing diversity of student populations may pose greater challenges for teachers in meeting the needs of all students, which could contribute to lower perceived teacher support and instructional clarity.

Additionally, changes in the curricula could be another factor that influences students' perceptions of teacher support and instructional clarity. The current curricula have become more demanding in terms of what is expected from students, and teachers are now expected to act as facilitators rather than giving strict directions (Carlgren et al., 2006). This shift in instructional practice places more responsibility on the students, which could lead to more students perceiving less teacher support. If teachers are not adequately trained or supported in providing effective instruction, students may perceive less support and instructional clarity in the classroom (Creemers et al., 2012; Darling-Hammond et al., 2017).

Although there is a lack of Icelandic studies investigating trends in teaching quality and assessment practice over time, prior research provides some insight into the prevalence of such practices. For example, studies from lower secondary schools indicate an overall low teaching quality in mathematics related to cognitive activation (Sigurjónsson, 2023) and instructional clarity related to feedback and clear learning goals (Svanbjörnsdóttir et al., 2023). While homework is generally considered an important practice by Icelandic school teachers, low-achieving students and their parents view it as too demanding, and these are the students that spend most time on homework (Sigurgeirsson & Björnsdóttir, 2016). This perception is unlikely unique to Iceland, and in what way such findings are related to achievement across diverse student groups needs to be investigated in future research.

The study also found an overall decline in homework frequency in mathematics and science between 2011 and 2019 across Nordic countries. This trend may be due to the growing emphasis on providing students with more meaningful and relevant learning experiences (Clement, 2010; Remmen & Iversen, 2022). Equity is another possible explanation for the decline in homework frequency in Nordic countries. Homework assignments may not always align with the needs and interests of all students, and low SES students may be less likely to have access to resources and parental support that can help them complete assignments effectively (Bempechat et al., 2011; Rønning, 2011). Consequently, some schools and educational systems may have shifted towards alternative forms of assignments that are more flexible and better aligned with students' backgrounds.

It is important to note that the reasons for the decline in the aspects of teacher support and instructional clarity, as well as homework frequency, are likely complex and multifaceted. Further research is necessary to identify the underlying factors contributing to these trends in order to inform policy and practice aimed at improving teaching quality and assessment practices in Nordic countries.

5.8.2 The Relationship Between Teaching Quality and Student Achievement

Classroom management was identified as having the largest correlation with student achievement in mathematics compared to other dimensions of teaching quality, which is consistent with previous research (e.g., Senden et al., 2023). Good classroom management is considered a prerequisite to facilitate other dimensions of teaching quality, such as in creating a supportive classroom climate and implementing cognitive activation (Charalambous & Praetorius, 2020), which can promote student achievement (Wolff et al., 2021). For example, by minimizing disruptions during learning, teachers can create a structured environment that fosters positive relationships between students and teachers and among students and maximize the amount of time available for cognitively challenging instruction.

This study also suggests that teacher support and instructional clarity seem to be better predictors of achievement at the student level compared to the classroom level. This could be due to individual differences in students' characteristics and background, such as their language abilities and socioeconomic status, which may impact the level and type of support they require from their teachers. Previous research has shown that students' perceptions of teaching quality can vary across diverse groups of students (e.g., Senden et al., 2023; Teig & Nilsen, 2022; Wang et al., 2018). Therefore, it is important to consider individual student characteristics when examining the relationships between teacher support, instructional clarity, and achievement. Measuring these factors at the student level may provide a more accurate understanding of these relationships.

Although previous studies have suggested that cognitive activation is an important dimension of teaching quality related to student learning outcomes (Baumert et al., 2010; Klieme et al., 2009; Lipowsky et al., 2009), the current study did not find any significant relationship between cognitive activation and achievement in mathematics and science across the Nordic countries. One possible explanation for this discrepancy could be related to the way cognitive activation was operationalized and measured in TIMSS, as some items were more related to low- rather than high-level of cognitive activation. For example, in mathematics, the items include memorizing rules, procedures, and facts. Therefore, this finding highlights the need for future research to explore the effects of various conceptualizations of cognitive activation and to what extent these variations matter for student achievement.

Measuring high-level cognitive activation can be challenging via the student or teacher questionnaire due to its context-specific nature. Thus, incorporating a qualitative perspective, like classroom observation, could be beneficial. For instance, a recent video study in Iceland reported frequent occurrences of low-level cognitive activation in mathematics classrooms (Sigurjónsson, 2023). Students often engaged in individual work that focused on procedural fluency, with limited connection to understanding mathematical concepts.

Additionally, the use of teacher rather than student questionnaires to measure cognitive activation may have impacted the findings. Teacher questionnaires may be more susceptible to social desirability bias, a tendency for teachers to answer the questionnaire in a way that will be perceived favorably by others (Muijs, 2006). By using the student questionnaires, this study found that the frequency of conducting experiments, as an indicator of cognitive activation in science, was related to student achievement in a non-linear pattern (inverted U-shape). This finding aligns with previous research (Cairns, 2019; Teig et al., 2018, 2021), and suggests that there may be an optimal level of conducting experiments that leads to the highest achievement in science.

Discrepancies in findings on the relationship between cognitive activation and achievement in mathematics and science may be due to how cognitive activation was measured and whether teacher or student questionnaires were used. Further research is needed to better understand this relationship and identify effective strategies for optimizing cognitive activation in these subjects.

5.8.3   The Relations Between Assessment Practice and Student Achievement

The findings of this study suggest that the associations between homework as part of teacher assessment practice and student achievement were limited. While the frequency of homework was positively related to mathematics achievement in Denmark and Sweden, homework time had a negative relationship with mathematics achievement in Finland. This finding may indicate that completing more homework does not necessarily lead to higher achievement, but rather that the quality and relevance of homework assignments are more important factors to consider (see a review by Fernández-Alonso & Muñiz, 2022). In Finland and Sweden, correcting assignments and providing feedback to students was negatively related to mathematics and science achievement, which is a finding that may warrant further investigation. These findings could indicate reversed causality, where low-achieving students spent more time completing homework and high-achieving students finished their work at school, and teachers placed more emphasis on correcting assignments and giving feedback to the struggling students. In Iceland, a recent study using classroom video data showed that although providing feedback is a common practice in mathematics classrooms, there was limited evidence of feedback being delivered with a clear purpose and of high quality to students (Svanbjörnsdóttir et al., 2023).

Moreover, no clear correlations emerged between the amount of emphasis teachers placed on various assessment strategies and student achievement in either mathematics or science. This may suggest that while assessment strategies are important for monitoring and evaluating student learning, they may not necessarily have a direct impact on student achievement. Furthermore, it is possible that the assessments used by teachers may not be aligned with the content and skills emphasized in the curriculum, and thus may not be as effective in promoting learning (Andrade & Brookhart, 2020; Gardner et al., 2010). Teachers may also need additional support and training in developing and using assessments that are aligned with the curriculum and promote student learning.

An alternative explanation might be linked to the limitations of the items utilized to measure assessment strategies in TIMSS. The items used in TIMSS might not encompass the complete array of assessment practices employed by teachers in the classroom, or they may not fully capture the complexity of these practices.

To conclude, this study highlights the need for a more holistic approach to teaching quality and assessment practices in mathematics and science education, one that considers multiple dimensions of teaching quality and emphasizes the need for further analysis of how the composition of the classroom may also impact the relationships between these constructs. With more nuanced understandings of these relationships, policymakers can allocate targeted resources, allowing educators to create a more supportive and equitable learning environment that promotes student achievement and success in mathematics and science.

Appendices

Appendix 1 Response Patterns of Teaching Quality Across Nordic Countries in TIMSS 2019

Classroom management (only in mathematics)

Student questionnaire: How often do these things happen in your mathematics lessons?

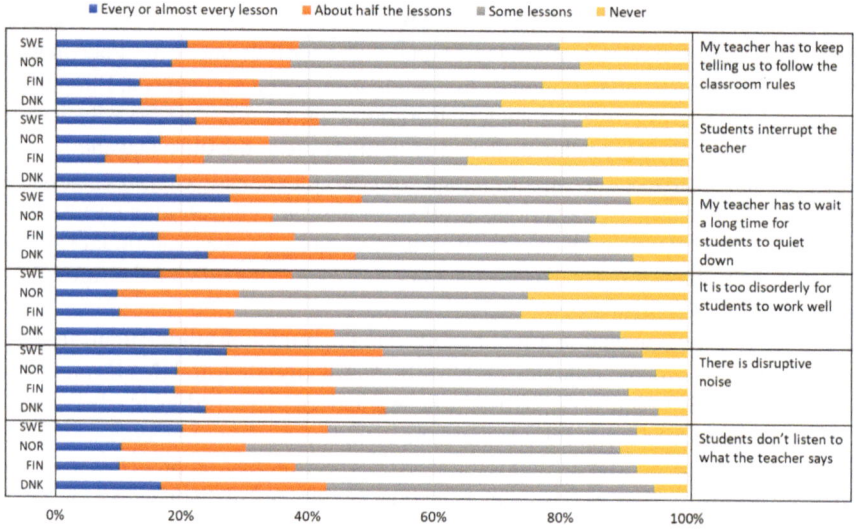

Teacher support and clarity of instruction in mathematics

Student questionnaire: How much do you agree with these statements about your mathematics lessons?

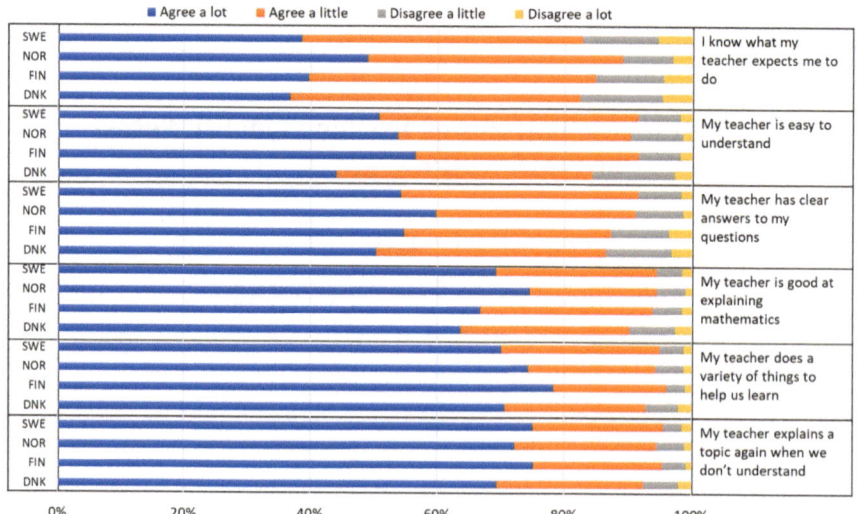

Teacher support and clarity of instruction in science

Student questionnaire: How much do you agree with these statements about your science lessons?

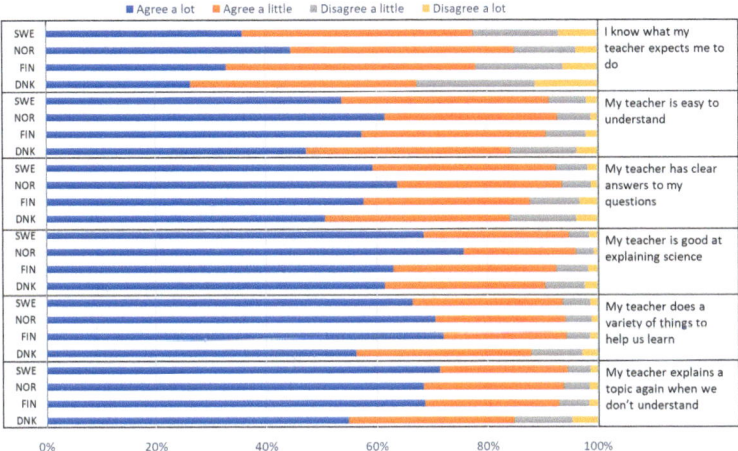

Cognitive activation (general)

Teacher questionnaire: In teaching mathematics/science to the students in this class, how often do you usually ask them to do the following?

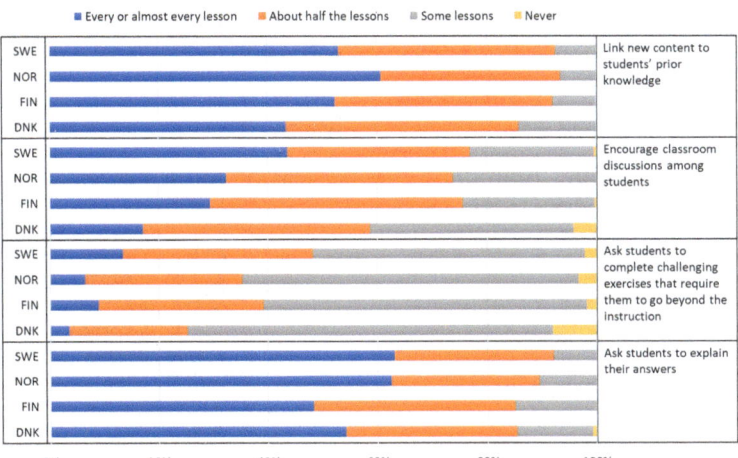

Cognitive activation in science

Student questionnaire: In science lessons, how often does your teacher ask you to conduct science experiments?

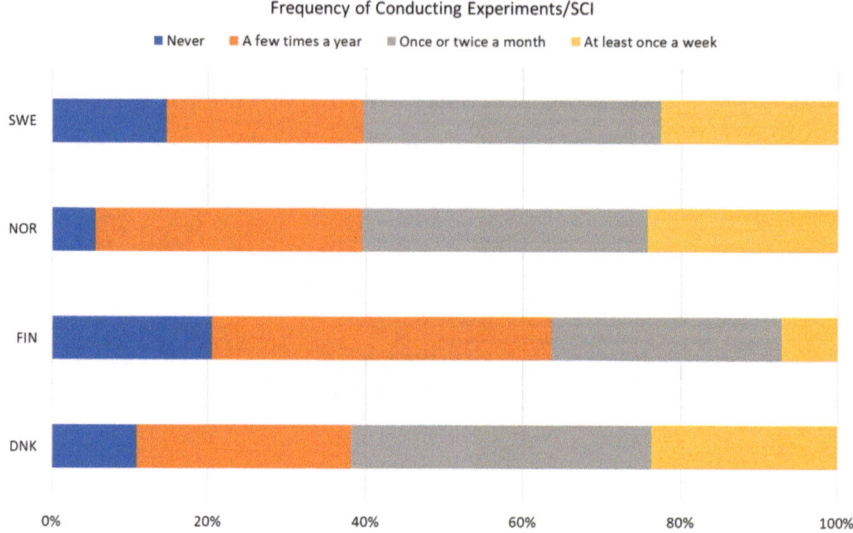

Appendix 2 Response Patterns of Assessment Practice Across Nordic Countries in TIMSS 2019

Homework frequency in mathematics and science

Teacher questionnaire: How often do you usually assign mathematics/science homework to the students in this class?

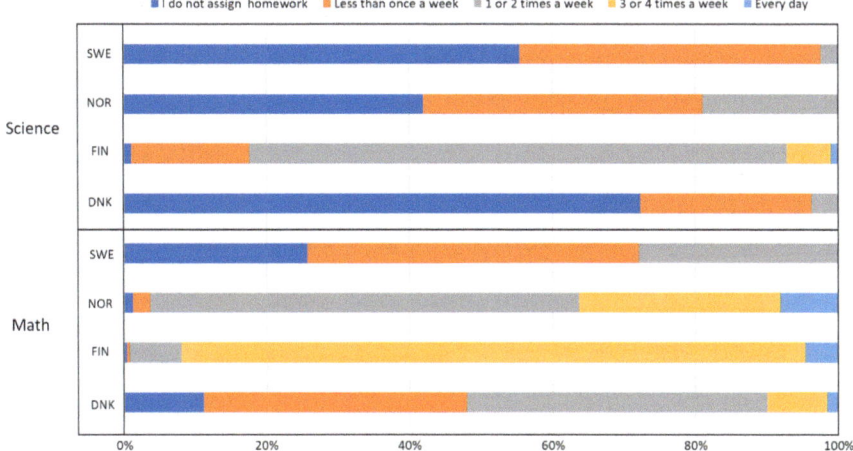

Duration of assigned homework in mathematics and science

Teacher questionnaire: When you assign mathematics homework to the students in this class, about how many minutes do you usually assign?

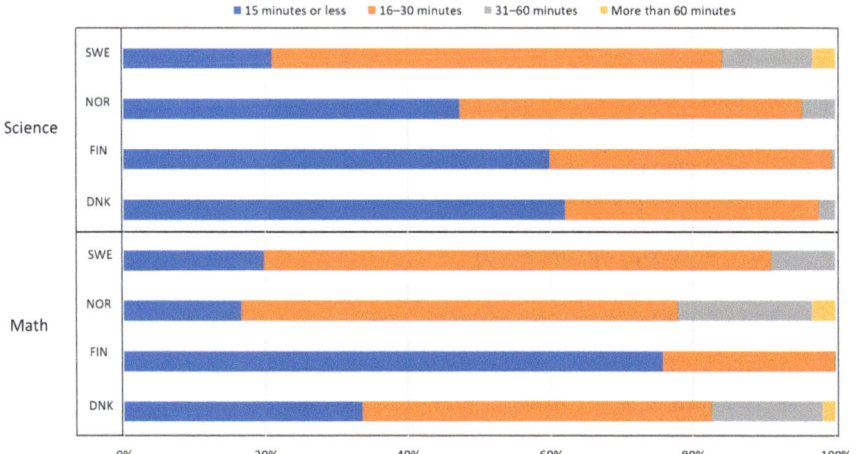

In-class homework discussion in mathematics

Teacher questionnaire: How often do you do the following with the mathematics homework assignments for this class?

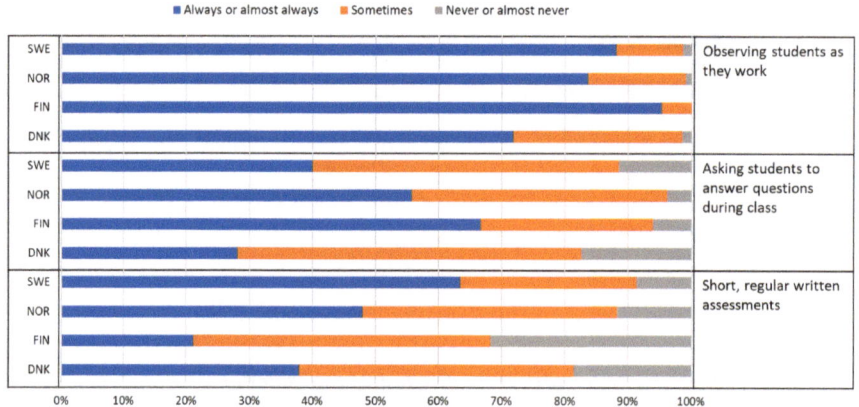

In-class homework discussion in science

Teacher questionnaire: How often do you do the following with the science homework assignments for this class?

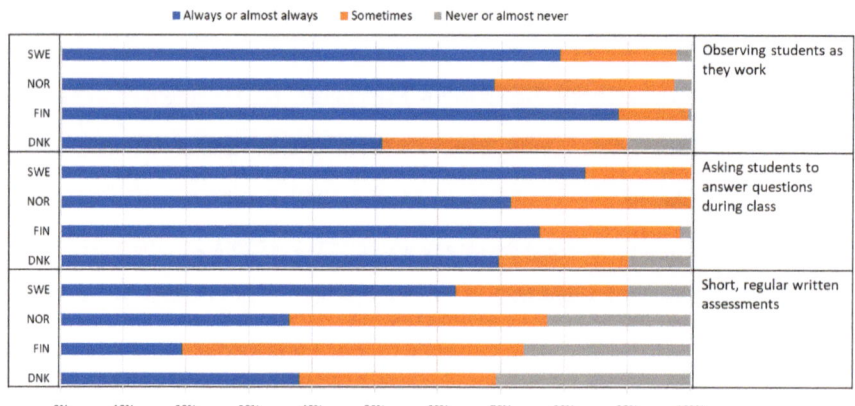

Teachers' emphasis on various assessment strategies in mathematics

Teacher questionnaire: How much emphasis do you place on the following sources to monitor students' progress in mathematics?

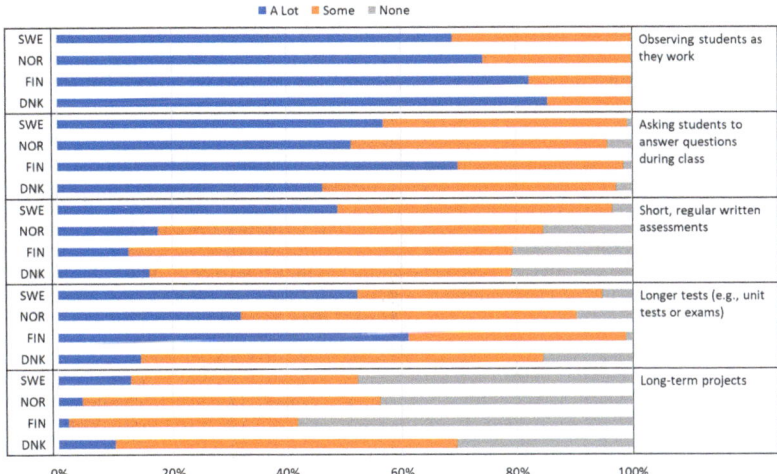

Teachers' emphasis on various assessment strategies in science

Teacher questionnaire: How much emphasis do you place on the following sources to monitor students' progress in mathematics?

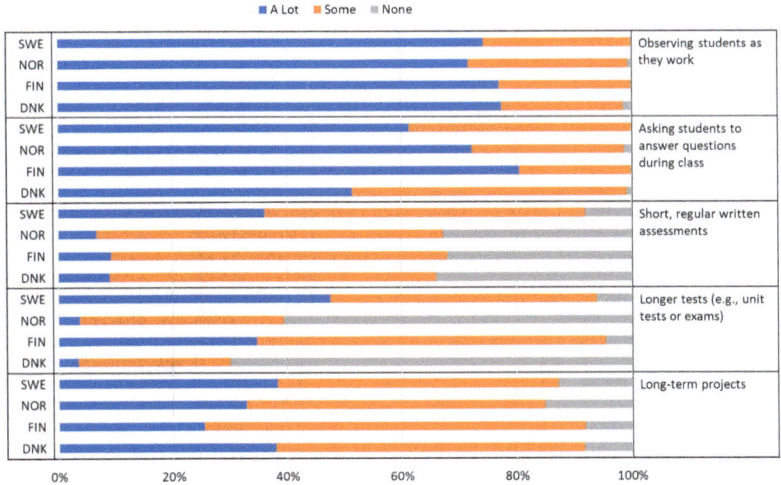

References

Andersson, C., & Palm, T. (2017). The impact of formative assessment on student achievement: A study of the effects of changes to classroom practice after a comprehensive professional development programme. *Learning and Instruction, 49*, 92–102. https://doi.org/10.1016/j.lea rninstruc.2016.12.006

Andrade, H. L. (2019). A critical review of research on student self-assessment. *Frontiers in Education, 4*. https://doi.org/10.3389/feduc.2019.00087

Andrade, H. L., & Brookhart, S. M. (2020). Classroom assessment as the co-regulation of learning. *Assessment in Education: Principles, Policy & Practice, 27*(4), 350–372. https://doi.org/10. 1080/0969594x.2019.1571992

Baumert, J., Kunter, M., Blum, W., Brunner, M., Voss, T., Jordan, A., Klusmann, U., Krauss, S., Neubrand, M., & Tsai, Y.-M. (2010). Teachers' mathematical knowledge, cognitive activation in the classroom, and student progress. *American Educational Research Journal, 47*(1), 133–180. https://doi.org/10.3102/0002831209345157

Bellens, K., Van Damme, J., Van Den Noortgate, W., Wendt, H., & Nilsen, T. (2019). Instructional quality: catalyst or pitfall in educational systems' aim for high achievement and equity? An answer based on multilevel SEM analyses of TIMSS 2015 data in Flanders (Belgium), Germany, and Norway. *Large-scale Assessments in Education, 7*. https://doi.org/10.1186/s40536-019-0069-2

Bempechat, J., Li, J., Neier, S. M., Gillis, C. A., & Holloway, S. D. (2011). The homework experience: Perceptions of low-income youth. *Journal of Advanced Academics, 22*(2), 250–278. https://doi.org/10.1177/1932202X1102200204

Black, P., & Wiliam, D. (2009). Developing the theory of formative assessment. *Educational Assessment, Evaluation and Accountability (formerly: Journal of personnel evaluation in education), 21*, 5–31. https://doi.org/10.1007/s11092-008-9068-5

Brown, T. A. (2015). *Confirmatory factor analysis for applied research*. Guilford Publications.

Cairns, D. (2019). Investigating the relationship between instructional practices and science achievement in an inquiry-based learning environment. *International Journal of Science Education, 41*(15), 2113–2135. https://doi.org/10.1080/09500693.2019.1660927

Carlgren, I., Klette, K., Mýrdal, S., Schnack, K., & Simola, H. (2006). Changes in Nordic teaching practices: From individualised teaching to the teaching of individuals. *Scandinavian Journal of Educational Research, 50*(3), 301–326.

Charalambous, C. Y., & Praetorius, A.-K. (2020). Creating a forum for researching teaching and its quality more synergistically. *Studies in Educational Evaluation, 67*. https://doi.org/10.1016/j.stueduc.2020.100894

Clement, N. (2010). Student wellbeing at school: The actualization of values in education. In T. Lovat, R. Toomey, & N. Clement (Eds.), *International research handbook on values education and student wellbeing* (pp. 37–62). Springer Netherlands. https://doi.org/10.1007/978-90-481-8675-4_3

Creemers, B., & Kyriakides, L. (2008). *The dynamics of educational effectiveness: A contribution to policy, practice and theory in contemporary schools*. Routledge.

Creemers, B., Kyriakides, L., & Antoniou, P. (2012). Teacher professional development for improving quality of teaching. *Springer*. https://doi.org/10.1007/978-94-007-5207-8

Darling-Hammond, L., Hyler, M. E., & Gardner, M. (2017). *Effective teacher professional development*. Learning Policy Institute.

Delahunty, T., Seery, N., & Lynch, R. (2020). Exploring problem conceptualization and performance in STEM problem solving contexts. *Instructional Science, 48*(4), 395–425. https://doi.org/10.1007/s11251-020-09515-4

Fan, H., Xu, J., Cai, Z., He, J., & Fan, X. (2017). Homework and students' achievement in math and science: A 30-year meta-analysis, 1986–2015. *Educational Research Review, 20*, 35–54. https://doi.org/10.1016/j.edurev.2016.11.003

Fernández-Alonso, R., & Muñiz, J. (2020). Homework: Facts and fiction. In T. Nilsen, A. Stancel-Piątak, & J.-E. Gustafsson (Eds.), *International handbook of comparative large-scale studies in education: Perspectives, methods and findings* (pp. 1–31). Springer International Publishing. https://doi.org/10.1007/978-3-030-38298-8_40-1

Fernández-Alonso, R., & Muñiz, J. (2022). Homework: Facts and fiction. In *International handbook of comparative large-scale studies in education* (pp. 1209–1239). Springer International Publishing. https://doi.org/10.1007/978-3-030-88178-8_40

Förtsch, C., Werner, S., Dorfner, T., von Kotzebue, L., & Neuhaus, B. J. (2017). Effects of cognitive activation in biology lessons on students' situational interest and achievement [journal article]. *Research in Science Education, 47*(3), 559–578. https://doi.org/10.1007/s11165-016-9517-y

Gardner, J., Harlen, W., Hayward, L., Stobart, G., & Montgomery, M. (2010). *Developing teacher assessment*. McGraw-Hill Education (UK).

Yang Hansen, K., Gustafsson, J.-E., & Rosen, M. (2014). School performance differences and policy variations in Finland, Norway and Sweden. In K. Yang Hansen, J.-E. Gustafsson, M. Rosen, S. Sulkunen, K. Nissinen, P. Kupari, R. F. Olafsson, J. K. Björnsson, L. S. Grønmo, L. Rønberg, J. Mejding, I. C. Borge, & A. Hole (Eds.), *Northern lights on TIMSS and PIRLS 2011: Differences and similarities in the Noordic countries* (pp. 23–45). Nordic Council of Ministers. https://doi.org/10.6027/TN2014-528

Hattie, J. (2009). *Visible learning: A synthesis of over 800 meta-analyses relating to achievement*. Routledge.

Herppich, S., Praetorius, A.-K., Förster, N., Glogger-Frey, I., Karst, K., Leutner, D., Behrmann, L., Böhmer, M., Ufer, S., Klug, J., Hetmanek, A., Ohle, A., Böhmer, I., Karing, C., Kaiser, J., & Südkamp, A. (2018). Teachers' assessment competence: Integrating knowledge-, process-, and product-oriented approaches into a competence-oriented conceptual model. *Teaching and Teacher Education, 76*, 181–193. https://doi.org/10.1016/j.tate.2017.12.001

Kanjee, A. (2009). Enhancing teacher assessment practices in South African schools: Evaluation of the assessment resource banks. *Education as Change, 13*(1), 73–89. https://doi.org/10.1080/16823200902940599

Klieme, E. (2019). *Teaching quality. Conceptualization, measurement, and findings for European countries*. Keynote lecture presented at the EU "PISA and Beyond Conference", https://doi.org/10.13140/RG.2.2.11701.55529

Klieme, E., & Nilsen, T. (2022). *Teaching quality and student outcomes in TIMSS and PISA* (Edited by T. Nilsen, A. Stancel-Piątak, & J.-E. Gustafsson, pp. 1089–1134). Springer International Publishing. https://doi.org/10.1007/978-3-030-88178-8_37

Klieme, E., Pauli, C., & Reusser, K. (2009). The pythagoras study: Investigating effects of teaching and learning in swiss and german mathematics classrooms. In T. Janik & T. Seidel (Eds.), *The power of video studies in investigating teaching and learning in the classroom* (pp. 137–160). Waxmann.

Krzywacki, H., Pehkonen, L., & Laine, A. (2016). Promoting mathematical thinking in Finnish mathematics education. In H. Niemi, A. Toom, & A. Kallioniemi (Eds.), *Miracles of education* (Vol. 2, pp. 109–123). Sense Publishers. https://doi.org/10.1007/978-94-6300-776-4_8

Lipowsky, F., Rakoczy, K., Pauli, C., Drollinger-Vetter, B., Klieme, E., & Reusser, K. (2009). Quality of geometry instruction and its short-term impact on students' understanding of the pythagorean theorem. *Learning and Instruction, 19*(6), 527–537. https://doi.org/10.1016/j.learninstruc.2008.11.001

Luoto, J. M., Klette, K., & Blikstad-Balas, M. (2022). Patterns of instruction in Finnish and Norwegian lower secondary mathematics classrooms. *Research in Comparative and International Education, 17*(3), 17454999221077848. https://doi.org/10.1177/17454999221077848

Marder, J., Thiel, F., & Göllner, R. (2023). Classroom management and students' mathematics achievement: The role of students' disruptive behavior and teacher classroom management. *Learning and Instruction, 86*, 101746. https://doi.org/10.1016/j.learninstruc.2023.101746

Marsh, H. W., Hau, K.-T., & Grayson, D. (2005). Goodness of fit evaluation in structural equation modeling. In A. Maydeu-Olivares & J. J. McArdle (Eds.), *Contemporary Psychometrics* (pp. 275–340). Lawrence Erlbaum.

Martin, M. O., Mullis, I. V., Foy, P., & Stanco, G. M. (2016). *TIMSS 2015 international results in science.* http://timssandpirls.bc.edu/timss2015/international-results/

Morin, A. J. S., Marsh, H. W., Nagengast, B., & Scalas, L. F. (2014). Doubly latent multilevel analyses of classroom climate: An illustration. *The Journal of Experimental Education, 82*(2), 143–167. https://doi.org/10.1080/00220973.2013.769412

Mostafa, T., Echazarra, A., & Guillou, H. (2018). *The science of teaching science: An exploration of science teaching practices in PISA 2015.* OECD Education Working Papers, No. 188, OECD Publishing. https://doi.org/10.1787/f5bd9e57-en

Muijs, D. (2006). Measuring teacher effectiveness: Some methodological reflections. *Educational Research and Evaluation, 12*(1), 53–74. https://doi.org/10.1080/13803610500392236

Mullis, I. V. S., Martin, M. O., Foy, P., Kelly, D. L., & Fishbein, B. (2020). *TIMSS 2019 international results in mathematics and science.* TIMSS & PIRLS International Study Center, Boston College. https://timssandpirls.bc.edu/timss2019/

Muthen, L. K., & Muthen, B. O. (2022). MPlus (Version 8.5) [Computer software]. Muthen & Muthen. https://www.statmodel.com/index.shtml.

Nilsen, T., Gustafsson, J.-E., & Blömeke, S. (2016). Conceptual framework and methodology of this report. In T. Nilsen & J.-E. Gustafsson (Eds.), *Teacher quality, instructional quality and student outcomes: Relationships across countries, cohorts and time* (pp. 1–19). Springer International Publishing. https://doi.org/10.1007/978-3-319-41252-8_1

Nilsen, T., Scherer, R., & Blömeke, S. (2018). The relation of science teachers' quality and instruction to student motivation and achievement in the 4th and 8th grade: A Nordic perspective In A. Wester (Ed.), *Northern lights on TIMSS and PISA 2018* (pp. 61–90). Nordic Council of Ministers. https://norden.diva-portal.org/smash/get/diva2:1237833/FULLTEXT01.pdf

Palm, T., Andersson, C., Boström, E., & Vingsle, C. (2017). A review of the impact of formative assessment on student achievement in mathematics. *Nordic Studies in Mathematics Education, 22*(3), 25–50.

Panadero, E., Jonsson, A., & Botella, J. (2017). Effects of self-assessment on self-regulated learning and self-efficacy: Four meta-analyses. *Educational Research Review, 22*, 74–98. https://doi.org/10.1016/j.edurev.2017.08.004

Praetorius, A.-K., Klieme, E., Herbert, B., & Pinger, P. (2018). Generic dimensions of teaching quality: The German framework of three basic dimensions. *ZDM Mathematics Education , 50*(3), 407–426. https://doi.org/10.1007/s11858-018-0918-4

Remmen, K. B., & Iversen, E. (2022). A scoping review of research on school-based outdoor education in the Nordic countries. *Journal of Adventure Education and Outdoor Learning*, 1–19. https://doi.org/10.1080/14729679.2022.2027796

Rohatgi, A., Hatlevik, O. E., & Björnsson, J. K. (2022). Supportive climates and science achievement in the Nordic countries: Lessons learned from the 2015 PISA study. *Large-Scale Assessments in Education, 10*(1), 12. https://doi.org/10.1186/s40536-022-00123-x

Rønning, M. (2011). Who benefits from homework assignments? *Economics of Education Review, 30*(1), 55–64. https://doi.org/10.1016/j.econedurev.2010.07.001

Ryu, E. (2014). Model fit evaluation in multilevel structural equation models. *Frontiers in Psychology, 5*, Article 81. https://doi.org/10.3389/fpsyg.2014.00081

Sass, D. A., & Schmitt, T. A. (2013). Testing measurement and structural invariance. In T. Teo (Ed.), *Handbook of quantitative methods for educational research* (pp. 315–345). SensePublishers. https://doi.org/10.1007/978-94-6209-404-8_15

Scheerens, J. (2016). *Educational effectiveness and ineffectiveness : A critical review of the knowledge base* (1st ed.). Springer. https://doi.org/10.1007/978-94-017-7459-8

Senden, B., Nilsen, T., & Teig, N. (2023). The validity of student ratings of teaching quality: Factorial structure, comparability, and the relation to achievement. *Studies in Educational Evaluation, 78*, 101274. https://doi.org/10.1016/j.stueduc.2023.101274

Sigurjónsson, J. Ö. (2023). *Quality in Icelandic mathematics teaching: Cognitive activation in mathematics lessons in a Nordic context*. University of Iceland.

Sigurgeirsson, I., & Björnsdóttir, A. (2016). Heimanám í íslenskum grunnskólum-Scope and vidhorf for students, parents and teachers [Homework in Icelandic Compulsory schools: Its Amount and Students', Parents' and Teachers' Attitudes]. Netla. https://ojs.hi.is/index.php/netla/article/view/2384

Sortkær, B. (2019). Feedback for everybody? Exploring the relationship between students' perceptions of feedback and students' socioeconomic status. *British Educational Research Journal, 45*(4), 717–735. https://doi.org/10.1002/berj.3522

Svanbjörnsdóttir, B., Zophoníasdóttir, S., & Gísladóttir, B. (2023). Quality of the stated purpose and the use of feedback in Icelandic lower-secondary classrooms results from a video study. *Teaching and Teacher Education, 121*, 103946. https://doi.org/10.1016/j.tate.2022.103946

Teig, N., Bergem, O. K., Nilsen, T., & Senden, B. (2021). Gir utforskende arbeidsmåter i naturfag bedre læringsutbytte? [Does inquiry-based teaching practice in science provide better learning outcomes?]. In T. Nilsen & H. Kaarstein (Eds.), *Med blikket mot naturfag [A view towards science]* (pp. 46–72). Universitetsforlaget.

Teig, N., & Nilsen, T. (2022). Profiles of instructional quality in primary and secondary education: Patterns, predictors, and relations to student achievement and motivation in science. *Studies in Educational Evaluation, 74*, 101170. https://doi.org/10.1016/j.stueduc.2022.101170

Teig, N., Scherer, R., & Nilsen, T. (2018). More isn't always better: The curvilinear relationship between inquiry-based teaching and student achievement in science. *Learning and Instruction, 56*, 20–29. https://doi.org/10.1016/j.learninstruc.2018.02.00

Teig, N., Scherer, R., & Nilsen, T. (2019). I know I can, but do I have the time? The role of teachers' self-efficacy and perceived time constraints in implementing cognitive-activation strategies in science. *Frontiers in psychology, 10*(1697). https://doi.org/10.3389/fpsyg.2019.01697

Teig, N., & Steinmann, I. (2023). Leveraging large-scale assessments for effective and equitable school practices: The case of the nordic countries. *Large-Scale Assessments in Education, 11*(21). https://doi.org/10.1186/s40536-023-00172-w

Wang, S., Rubie-Davies, C. M., & Meissel, K. (2018). A systematic review of the teacher expectation literature over the past 30 years. *Educational Research and Evaluation, 24*(3–5), 124–179. https://doi.org/10.1080/13803611.2018.1548798

Wolff, C. E., Jarodzka, H., & Boshuizen, H. P. A. (2021). Classroom management scripts: A theoretical model contrasting expert and novice teachers' knowledge and awareness of classroom events. *Educational Psychology Review, 33*(1), 131–148. https://doi.org/10.1007/s10648-020-09542-0

Nani Teig is an associate professor at the University of Oslo, Norway. Her research focuses on inquiry-based teaching, scientific reasoning, teaching quality, and academic resilience. She integrates multilevel analyses using data from videos, surveys, assessments, and computer log files. Dr. Teig has received several awards and fellowships, including the Global Education Award, Bruce H. Choppin Dissertation Award, Young CAS Fellow, and UNESCO GEM Fellow.

Jennifer Maria Luoto (born 1988) is a Postdoctoral fellow at the Department of Teacher Education and School Research, University of Oslo. Her research focuses mainly on classroom observations and teaching quality in different classrooms and national contexts.

Chapter 6
Are Changes in Content Coverage Related to Changes in Achievement Over Time?

Monica Rosén⊙ **and Trude Nilsen**⊙

6.1 Introduction

Understanding changes in educational achievement over time is an essential goal for most stakeholders within the field of education. Countries taking part in recurring International large-scale assessments, such as IEA's Trends in International Mathematics and Science Study (TIMSS), are not only provided with reliable and comparable measures of achievement trends in mathematics and science but also with a vast amount of contextual data, which are valuable for further analysis. Educational circumstances and potential causes behind changes in educational outcomes are at the core of interest for policymakers, practitioners, researchers, students, parents, and the public. It is often difficult to provide evidence of the reasons behind changes in achievement. Not only because measures of knowledge, for example, the TIMSS achievement scores in mathematics, are complex, in the sense that there are very many determining factors behind student performances on these tests, but also because the design of these studies is cross-sectional, and this makes it hard to determine what is the cause and what is the result, as a cause must appear before the effect. However, there are statistical techniques and methodologies with which one can investigate theories of plausible causes despite the cross-sectional design (see Gustafsson &

M. Rosén (✉)
Department of Education and Special Education, University of Gothenburg, PO BOX 300, SE 405 30 Gothenburg, Sweden
e-mail: Monica.Rosen@ped.gu.se

T. Nilsen
Department of Teacher Education and School Research, CREATE—Centre for Research on Equality in Education (project number 331640), University of Oslo, Postboks 1099, Blindern, 0317 Oslo, Norway
e-mail: trude.nilsen@ils.uio.no

© The Author(s) 2024
N. Teig et al. (eds.), *Effective and Equitable Teacher Practice in Mathematics and Science Education*, IEA Research for Education 14,
https://doi.org/10.1007/978-3-031-49580-9_6

Nilsen, 2022, for an overview and examples). Even if each cycle of TIMSS is cross-sectional for the respondents, it is longitudinal at the country level. This fact opens possibilities to investigate factors that may have contributed to the observed changes in achievement.

In this chapter, we focus on the classroom level, and investigate changes in students' Opportunity to Learn (OTL) with respect to the core aspect of OTL at the implemented level of the curriculum, content coverage. We investigate if changes in content coverage may account for any of the changes in mathematics and science achievement in grade four over the three TIMSS cycles 2011, 2015, 2019 in the Nordic countries. The concept content coverage refers to the content assessed by the TIMSS achievement tests in mathematics and science, as reported by the teachers of the assessed students who reported whether the topics in the test had been taught to the class before the tests were administered.

6.2   Theoretical Framework

In this chapter we address OTL at the implemented level of the curriculum, i.e., the classroom level, to find out if students in schools have had similar content coverage in school as is represented in the TIMSS tests. The question relates to the concept of OTL, which is an old multidimensional educational construct with roots in curricular theory (Dahllöf, 1970; Husén & Dahllöf, 1965) and also in learning and instruction research (Caroll, 1963) and further developed within the framework of educational effectiveness research (e.g., Scheerens, 2016). The curricular strand of OTL was coined by IEA (International Association for the Evaluation of Educational Achievement) in the 1960s when the research design for the first early international large-scale assessments of mathematics and science achievements was developed (Husen, 1967; Pelgrum, 1989). Students' OTL is one of those multidimensional factors that hold explanatory power at both school- and system levels (The OTL construct and its relation to different curricular levels are explained in Chap. 4).

Measuring students' OTL is challenging, and its measures vary over the years (see e.g., Schmidt et al., 2009; McDonnell, 1995; for a brief history of the construct's origin and development). Nevertheless, OTL measures are taken in TIMSS, which aim to indicate whether the students have had an opportunity to learn the topics or cognitive tasks included in the achievement test.

The aim of measuring OTL is to reflect what opportunities have been offered in the classroom to learn the tasks and content of the test and is primarily aligned with the implemented curriculum. In the classroom, the following three teaching factors moderate students' opportunities to perform on knowledge and skills tests: content coverage, the time spent on the content/task and quality of teaching. *Content coverage* refers to whether the tasks and content of the test have been taught to the student or not, and to what extent. *Time* on content refers to how much time has been spent learning the content/tasks. The time spent on the task also includes *timing,* which refers to the time gap between when the content was taught and the test administration. *Quality of teaching* refers to the teacher's competence to teach the content to the students and provide a good learning environment for the students. Of these three OTL factors, content coverage is the first and necessary condition for enabling students to achieve on the test (Schmidt et al., 2010). Time on the task or content, and quality of teaching, come second and are relevant moderating factors to determine the level of learning.

These days, the OTL construct is often used in a much broader sense, including factors related to students' backgrounds and the learning conditions outside of school. For IEA, the measure of OTL was, and still is, primarily meant to capture curricular differences when studying achievement differences across school systems. Indicators of OTL are included in all IEA studies and observed at different curricular levels. Chapter 4 offers a comprehensive presentation of the links between different curricular levels and the OTL construct, and a presentation of the indicators used in TIMSS.

In this chapter, the investigation of OTL factors' relationship with achievement and changes in achievement is limited to content coverage at the classroom level and whether the student has been taught the topics included in the TIMSS assessment of student achievement. Differences in content coverage across classrooms and time may account for some of the differences and changes in achievement.

6.2.1 Content Coverage in Grade Four Mathematics

Relative to other predictors of achievement, at present, there are fewer studies on the relationships between achievement on large-scale international assessments and content coverage in the classroom. However, one of relevance is a study by Scheerens (2016), where the relationship between OTL and achievement was investigated in both TIMSS 2011 (grade four and grade eight) and PISA 2011 (15-year-olds). Scheerens conducted a series of regression analyses to assess the effect of OTL on mathematics and science achievement while controlling for the "books at home"-variable as a proxy for students' home background. The many content coverage items from the teacher questionnaire in TIMSS were combined first into domain indices (three in science and three in mathematics), all with a scale from 0 to 100, and then averaged into a single mathematics OTL index and a single science OTL index. The findings for grade four were surprising. The results showed that content coverage in mathematics was significantly related to mathematics achievement in only 12 out of the 23 included countries, and the average effect was a modest 0.074. In science, there were virtually no effects of the science OTL index on achievement. Scheerens concludes that there are methodological challenges attached to the OTL indicators and calls for a closer examination of the validity of the OTL measures in TIMSS.

6.3   Research Question

The overarching aim of the present study is to investigate *whether changes in content coverage are related to the changes in mathematics and science achievement from 2011 to 2019 in Sweden, Norway, Finland, and Denmark.* We address this aim more specifically through the following research questions:

1. *Is there a positive correlation between the OTL measures of content coverage and achievement?*
2. *Has the amount of content coverage changed from 2011 to 2019?*
3. *Are the possible changes in content coverage related to changes in grade four mathematics and science achievement from 2011 to 2019?*
4. *Are there notable differences between Denmark, Finland, Norway, and Sweden?*

6.4   Methodology

6.4.1   Data and Sample

In the present study, we use data from the 2011, 2015, and 2019 TIMSS cycles. We include data from the teacher questionnaires and the results from the students' achievement tests in mathematics and science. Included are the fourth graders and their teachers from the Nordic countries that participated in these cycles: Sweden, Norway, Finland, and Denmark. In Norway, however, the target population changed in 2015 from fourth to fifth grade, so our sample includes Norwegian fourth graders in 2011 and fifth graders in 2015 and 2019. Findings from Norway hence need to be interpreted with caution. Part of the explanation for the large increase in achievement in Norway from 2011 to 2019 is that students are both older and have one more year of schooling (Olsen & Bjørnsson, 2018). For further descriptions of TIMSS data, including sampling, plausible values, and weights, please see Chap. 3.

6.4.2   Measures

In the present study, we use a set of derived variables available in the TIMSS database which are based on teachers' responses to whether and when they have covered the content within each of the three content domains for mathematics (number, geometry, and data) and the three domains for science (life science, physical science, and earth science). There are a number of different topics within each domain for which the teacher is asked to select one of the following three response options: "mostly taught before this year", "mostly taught this year", "not yet taught or just introduced". For example, within the domain number, one topic (out of many) is: "adding, subtracting, multiplying, and dividing with whole numbers". It should be noted that these content coverage questions do not include any information about to what extent the topics have been taught. These topic items were then combined to form indices, one for each domain in mathematics and science, respectively.

The description of how TIMSS constructed the indices is available in the TIMSS 2019 User Guide for the International Database (Fishbein et al., 2021). In short, the three response options were re-coded into two categories, to inform whether the topics had been taught or not. After that, the questionnaire items were combined into indices, one for each content domain. The indices then reflect the percentage of topics within each content domain (three in mathematics and three in science) that had been taught to the students.

In mathematics, the three indices represent *percentage of number topics taught, percentage of measurement and geometry topics taught,* and *percentage of data topics taught.* For simplicity, in this study, we refer to these content coverage indices as *OTL in number, OTL in geometry,* and *OTL in data,* respectively.

Similarly, for science, the three content coverage indices in the TIMSS database represent the *percentage of life science topics taught, percentage of physical science topics taught,* and *percentage of earth science topics taught,* and we refer to these indices as *OTL in life science, OTL in physical science,* and *OTL in earth science.*

Descriptive information for the OTL variables in mathematics, in the three TIMSS cycles studies and for all four countries analyzed, is presented in the bar graphs below (Fig. 6.1). Additional descriptive statistics for these OTL mathematics indices are presented in Chap. 4, along with those for science. The descriptive information about the OTL variables is selected from the TIMSS 2019 International Results in Mathematics and Science (Mullis et al., 2020).

The graphs show that the average proportion of topics covered in each mathematics content domain is relatively high, but also that it changes across time in all countries. The graphs also show that the countries differ in the pattern of change.

6.4.3   The Analytical Approach

The method of analyses in this study resembles that of longitudinal growth models (Murnane & Willete, 2010) but is adjusted to trend analyses. Such causal methods enhance the robustness of inferences (Gustafsson, 2010).

Causality and causal language. To investigate the relationship between predictors and outcomes in ILSA, most studies utilize data from one cycle only (Scherer, 2022). Some include two or more cycles and do the analyses separately for each cycle and compare results across time. The present study merged the data from three cycles (2011, 2015, and 2019) and used a causal method to utilize the trend design of TIMSS (see Chap. 2 for details on the trend design of TIMSS). This approach is far more robust and enhances the plausibility of causal inferences, as well as the reliability and validity of inferences.

The analytical approach. The analytical approach in the present chapter is the same as in Chap. 7, a structural equation model (SEM) with mediation. We used the

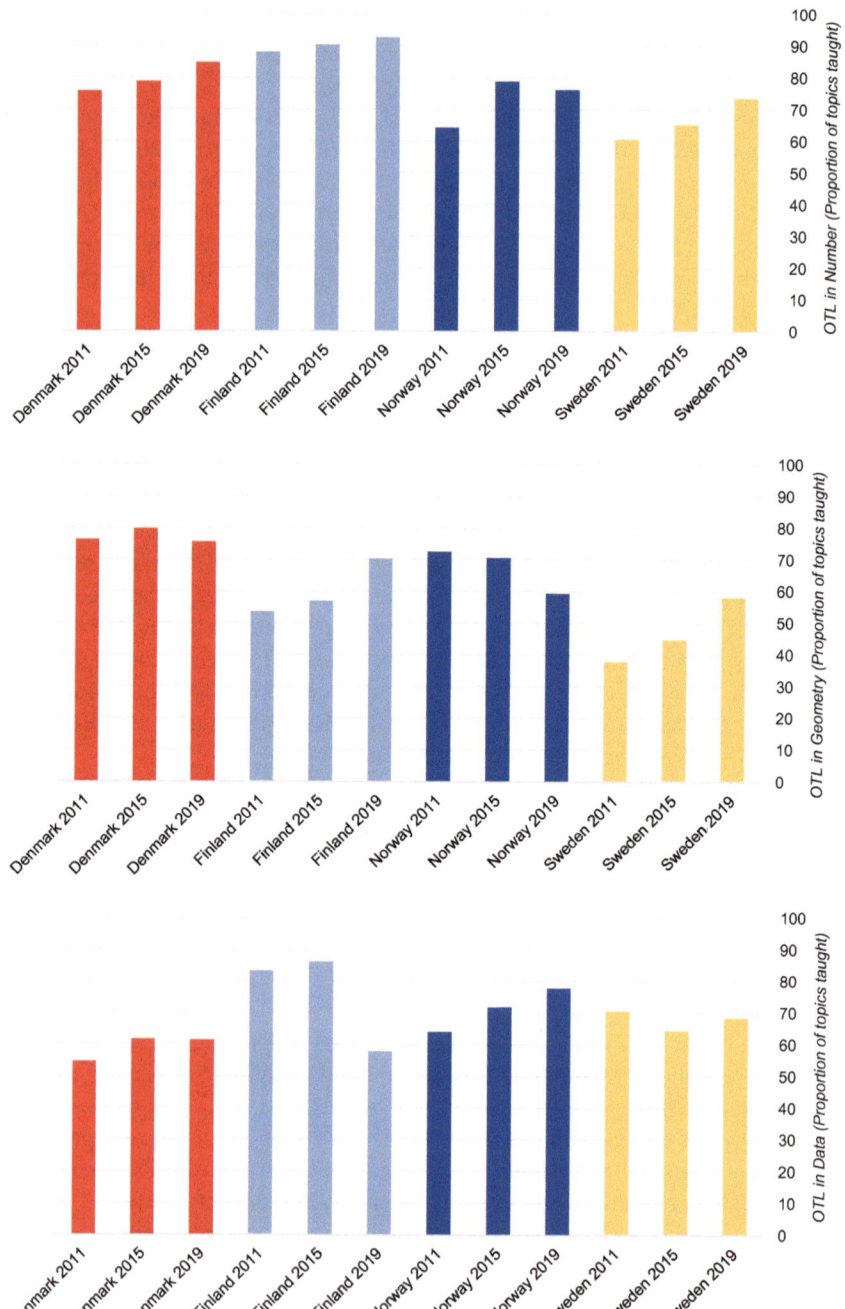

Fig. 6.1 Bar graphs showing the average proportion of topics in the TIMSS mathematics test for grade four that had been taught to the students at school by the time of testing. *Note* One bar per country and cycle for each content domain. Data derived from the teacher questionnaire

Fig. 6.2 The null model

software Mplus 8 for the analyses (Muthén & Muthén, 2017). Student and teacher data were merged using the IEA International Database Analyzer (IDB Analyzer), and a dummy variable called *Time* was added to each of the three datasets for the three cycles. *Time* is coded 0 for 2011, 1 for 2015, and 2 for 2019. The data for the three cycles were then merged. In Mplus, a null model estimate changes in achievement over time. In Fig. 6.2, *c* is the regression coefficient for the relation between time and achievement and describes the slope of the effect of time on achievement. The c-coefficient represents the average change in achievement for one unit change in time. Thus, to get the average change in achievement between 2011 and 2019, the c-coefficient should be multiplied by two.

We hypothesize that other factors besides the passing of time may "explain" changes in achievement. More specifically, we hypothesize that changes in OTL may account for any of the changes in achievement. Another (more technical) way of phrasing this is that we hypothesize that OTL to some degree may *mediate* the effect of time on achievement, as illustrated in Fig. 6.3.

Suppose OTL mediates the effect of time on achievement. In that case, it may mean that changes in OTL are related to changes in achievement. It may, in turn, indicate that changes in OTL, which in our analysis refers to changes in content coverage in the classrooms, explain changes in achievement over time.

If OTL has improved over time, the regression coefficient **a** would be positive. If OTL is positively related to achievement, the regression coefficient **b** is positive. However, we still needed to test whether the mediation is significant, which was done through the command *Model indirect* in Mplus. This estimate of the indirect effect is the main focus of the present study, as it tells us whether, and to what extent, OTL may mediate (or "explain") changes in achievement.

This analysis aims to explain *student achievement changes over time,* not differences in achievement between classrooms. Therefore, the analyses were done at

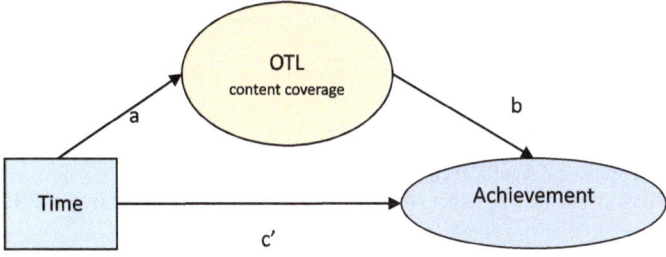

Fig. 6.3 Hypothesized mediation model in which OTL mediates the effect of time on achievement

the student level. However, to take into account the hierarchical design of the data where students are nested within classes that are nested within schools, and to avoid under-estimation of standard error, we used the option in Mplus called "TYPE = COMPLEX" where the data is clustered at the class-level. This way, the analyses take the hierarchical clustering of students and the between classroom variation into account.

6.5 Results

6.5.1 The Base Model—Changes in Achievement Over Time

We started by examining the effect of time on achievement. The results reflect the slope of the relation between time and achievement. A positive regression coefficient means that achievement increased over time. A large positive coefficient would reflect a large increase. The regression coefficients reflect the mean change in achievement across the three cycles and are provided in Table 6.1.

The findings show an average increase in science and mathematics achievement from 2011 to 2019 for Norway and Sweden. The numbers represent changes in score points on the test from one time-point to the next. The total change in achievement from 2011 to 2015 to 2019 is thus 2 * the regression coefficient, which for Sweden equals 10.6 score points in science and 22.6 score points in mathematics. For Norway, a substantial part of the large increase (2 * 22.6 = 45.2 score points in mathematics and 2 * 22.1 = 44.2 score points in science) is explained by the shift of target grade in Norway from grade four to five in 2015 (Olsen & Bjørnsson, 2018). For Finland and Denmark, the achievements declined in both subject domains. The regression model assumes a linear relationship between time and achievement, that is that the changes are evenly distributed between study cycles, which is not necessarily the case as can be noted when looking at the descriptive statistics for each time point. Instead, these model estimates represent the average across the three cycles of TIMSS analyzed in this study. For further details on changes in achievement, see Chap. 1.

Table 6.1 Regression coefficients for the effect of time on achievement in mathematics and science in grade four

Subject domain	SWE	NOR[a]	FIN	DEN
Science	5.3*	22.1*	− 6.3*	− 3.1*
Mathematics	11.3*	22.6*	− 6.1*	− 6.1*

Note * Indicate statistical significance at the $p < 0.05$ level
[a] Note that Norway changed its target population from grade four to grade five in 2015

6.5.2 Relations Between Changes in OTL and Changes in Mathematics Achievements

Interpretation of results. For the analyses of the percentage coverage of the topic within each content domain in mathematics, we first analyzed each domain separately, and these results are shown in Table 6.2. This table presents the results as text and symbols (see Appendix 1 for all estimates). To interpret the results, one must remember that previous research and our hypothesis predict that OTL should be positively related to achievement and that if OTL increases over time, it should cause increased achievements. If OTL decreases over time, it should cause a decline in achievement.

However, the four Nordic countries analyzed in this study have different achievement profiles over time (see Chap. 1). In Norway and Sweden, achievements have increased from 2011 to 2019, while in Denmark and Finland, achievements decreased. Here, we use Finland as an example to illustrate how to interpret results. Firstly, one must recall that Finland had declining student achievements from 2011 to 2019. If OTL is positively related to achievement and if OTL increases over time in Finland, and if the indirect effect is significant and positive, it means that OTL probably prevented a further decline in Finland's negative trend in achievement. Hence, an increase in OTL over time could prevent further achievement declines for countries with a negative achievement trend. For countries with positive achievement trends, an increase in OTL may explain part of this increase. At the same time, a

Table 6.2 How changes in the proportion of students having received instruction in the topics; number, geometry, and data are related to changes in mathematics achievement over time

	What is the average effect of content coverage (OTL) on overall mathematics achievement?				How has content coverage changed over time? (has it decreased ↓, or increased ↑)				How are changes in content coverage related to changes in achievement (indirect effect in score points)?			
	SWE	NOR	FIN	DEN	SWE	NOR	FIN	DEN	SWE	NOR	FIN	DEN
Content coverage number topics	NS	Small, positive effect	Small, positive effect	Small, positive effect	↑	↑	↑	↑	NS	↑ by 3	↑ by 1	↑ by 1
Content coverage geometry topics	NS	Small, positive effect	Small, positive effect	Small, positive effect	↑	↓	↑	NS	NS	↓ by 2	↑ by 1	NS
Content coverage data topics	NS	Small, positive effect	Small, positive effect	Small, positive effect	↓	↑	↓	NS	NS	↑ by 1	↓ by 1	NS

Note ↓ denotes a decrease, ↑ denotes an increase. OTL means the opportunity to have learned the test content in school

decrease in OTL may have prevented an additional increase in the already positive achievement trend.

Results. We first provide the results for the relations between the different OTL measures and mathematics achievement, which are provided in the first column of Table 6.2 ("what is the effect of OTL on mathematics achievement?"). There were no significant relations between any OTL variables and mathematics achievement for Sweden, but slight, positive, and significant findings for all the other countries for all three OTL measures. Minor effects here reflect effect sizes below 0.2. In other words, a clear pattern of positive associations between OTL in mathematics and achievement exists.

OTL in number increased from 2011 to 2019 in all the four Nordic countries. This means teachers reported higher percentages of students having covered the topics in the content domain number in 2019 compared to 2011. In Norway, this increase in OTL for the content domain number, accounts for about 6.5 points of their 45 points increase in achievement from 2011 to 2019. The remaining 38 points of increase are explained by other factors (e.g., the change in target grade from grade four to five and the age of students). Achievements in both Finland and Denmark decreased from 2011 to 2019; however, had it not been for the increase in OTL for the content domain number, their achievements could have decreased by yet another two points. In Sweden, the OTL for the content domain number was not significantly related to achievement; thus, it could not explain any part of the increased achievements in Sweden.

OTL in geometry increased from 2011 to 2019 in Sweden and Finland but decreased in Norway. There was no significant change in Denmark. OTL in geometry was related to changes in achievement only in Norway and Finland. In Norway, the decrease in OTL in geometry, hindered a further increase in achievements. The indirect effect was about minus 3.3 points, meaning that Norway's achievement could have increased by about 3 points, had it not been for this decrease in OTL. In Finland, the opposite was the case. In Finland, mathematics achievement decreased by 12 points from 2011 to 2019. Had it not been for the increase in OTL in geometry, their achievements could have declined by almost 14 points.

OTL in data did not change significantly in Denmark, but in Sweden and Finland the level decreased, and in Norway, it increased. The increase in OTL in data in Norway explained two of the 45 points of increased achievement over time (and/or from grade four to grade five). In Finland, the indirect effect was about two points. This means that had it not been for the decrease of OTL in data over time, their achievements could have declined by only 10 points rather than 12 points.

The results from the analyses of the mediation model, where *all three OTL variables* were included as mediators at the same time, are provided in Table 6.3. The relations between OTL and achievements and changes in OTL over time are not different from the results provided in Table 6.2. Hence, only the indirect effects for each OTL variable and the total indirect effect are provided. The total indirect effect reflects the sum or the total contribution of the three OTL variables when they are controlled for each other.

Table 6.3 Indirect effects and the total indirect effect for the model where all three OTL variables in mathematics were included simultaneously in one model

	SWE	NOR	FIN	DEN
Indirect effect. OTL number	0.7	3.0^{**}	0.6^{**}	0.8^{*}
Indirect effect. OTL geometry	0.5	-1.8^{*}	0.8^{**}	0.0
Indirect effect. OTL data	0.0	1.9^{*}	-0.2	0.3
Total indirect effect. All OTL variables included	1.3	2.2^{**}	1.2^{*}	1.0^{*}

Note Indicate statistical significance at the * $p < 0.05$ level, and ** at the $p < 0.01$ level. Model fit was good

Similar to the results provided in Table 6.2, there were no significant effects for Sweden. For Norway, the total indirect effect of all three OTL variables was about four points. Meaning that four out of the 45 points of increase in achievement may be explained by changes in OTL. Note that the increase in achievement in Norway may stem from changes in time and/or the change in grade (from grade four to five). However, while OTL in number and data contributes positively, OTL in geometry contributed negatively. The total effect for Finland was about one point, which reflects a positive contribution by OTL. Had it not been for the increase in OTL in geometry and number, their achievements could have declined by another two points. In Finland, there was no significant contribution from OTL in data. In Denmark, only OTL in number contributed significantly to the total indirect effect. Like Finland, Denmark's decline in achievement could have been worse had it not been for the increased OTL in number.

6.5.3   Relations Between Changes in OTL and Changes in Science Achievements

For science, none of the relations between the three OTL variables (OTL in life science, OTL in physical science, and OTL in earth science) and achievement were significant. This may be because the proportion of missing data for these OTL variables was relatively high in all countries but Finland, so the results may not be trusted. Appendix 2 presents the proportion of teachers/classrooms missing information on the OTL variables for all countries in each of the analyzed TIMSS cycles.

Due to the absences of data in the OTL variables, we did not include a large table like Table 6.2 for science achievement. The mediation model indicated that neither the indirect effect of each OTL variable nor the total indirect effects of the three OTL variables joined together were significant for any country (see Appendix 3). Regarding significant changes, there was a large decline in OTL for earth science over time and a slight decline in life science in Norway. Swedish data also indicated a significant decline concerning OTL in earth science. There were no significant changes over time in Finland or Denmark. However, whether these results may be generalized is doubtful as the level of missing responses to the content coverage questions in the teacher questionnaire were high in Denmark, Norway, and Sweden.

6.6 Summary and Discussion

This study aimed to investigate if changes in students' opportunities to learn the content and tasks included in the TIMSS tests can explain any of the changes in achievement. The first question was if the OTL measures of content coverage were positively related to achievement. In mathematics, the content coverage items are combined into three OTL-scales for the three different mathematical content domains in TIMSS; one for number, one for geometry and measures, and one for data. The items in each scale were aimed to indicate whether the topics within the domain had been taught to the students. All three OTL scales in mathematics were positively related to mathematics achievement on the tests in all four countries, The effect was small and statistically significant in all countries except Sweden, where the effect was too small to be significant.

The OTL scales in science were constructed in a similar way to mathematics, one for each content domain in the science TIMSS test; one for life science, one for physical science and one for earth science. However, the proportion of missing data on these scales was too high in Norway, Sweden, and Denmark for any reliable interpretation of the analysis of relationships. Finland had an acceptable level of missing data, but since no reliable comparisons could be done with the other Nordic countries, we chose to not discuss those results.

The second question was on whether there had been any change between 2011 and 2019 in the OTL measures. An increase was found in OTL in the number domain in all countries. An increase in OTL in geometry was found in Sweden and Finland, while Norway showed a decrease and Denmark showed no change. OTL in the data domain decreased in Sweden and Finland but increased in Norway. There was no change in Denmark. Finland and Denmark reported higher levels of content coverage in the number content domain compared to Norway and Sweden. Finland also reported higher levels of content coverage in data, but apart from this, the differences in content coverage between the Nordic countries were small. The overall picture indicates a medium–high level of OTL in all three mathematics content areas for all Nordic countries.

The last question was on whether the changes in content coverage explain any achievement changes. The analysis showed small but significant contributions of all three OTL variables in Norway and Finland. In Denmark, only the change in OTL in the number domain contributed to the change in achievement, whilst in Sweden, none of the changes in OTL (mathematics domains) are related to achievement.

We can conclude that our results align with previous findings in mathematics; more content coverage is positively associated with higher achievements in all countries (Scheerens, 2016). Furthermore, an increase in content coverage over time is associated with increased achievements (or a less negative decline), while a decrease from 2011 to 2019 in students' learning opportunities is associated with decreased achievements (or a less positive incline in achievement).

It must be pointed out that although most findings were significant, the effects are minor. There may be several reasons for this, one is that average level content coverage is high in all countries. Moreover, the changes in content coverage appear small and mostly in the positive direction. Low levels of variation cause low effects. For policymakers and teachers, low effects of content coverage are desired as it signals that the assessed topics have been addressed in most of the classrooms.

Another reason for the small effects could be related to the reliability of the OTL scales, which are low due to limited items and response options. Furthermore, the questions in the teacher questionnaire regarding topics covered in the classroom can be interpreted in different ways, potentially adding some inconsistency or noise to these measures. Another reason for minor effects may be the small sample of teachers participating, which affects the power. Had similar questions been asked to the representative samples of students, the effects would probably have been larger (see e.g., Schmidt et al., 2011; Scheerens, 2016). However, it would be hard for fourth-grade students to answer such questions.

Finally, it should be noted that all the content coverage items in the teacher questionnaire suffer from large levels of non-responses, in science, even more than in mathematics. This is true for Norway, Denmark, and Sweden. This is unfortunate as it also contributes to weakening the effect sizes. No such missing problems were found in the Finnish data.

6.6.1 Limitations, Reliability, and Validity

The method used in the present study is more robust than analyses of one cycle of TIMSS, and more robust than comparing results from separate analyses of each cycle. Data is longitudinal at the country level, which is good when investigating the effect of system level factors, as many plausible important factors (social, cultural, and economic) remain stable over time at this level. Nevertheless, in this analysis, we only investigate a limited number of potential country level factors that may underlie the achievement changes between 2007 and 2019. So strictly speaking, no causal inferences can be made regarding content coverage as an explanatory factor to changes in grade four mathematics achievement. We can, however, conclude that our results concord with previous research and theory, that content coverage matters, and on average, when content coverage increases, so does achievement.

One limitation of this study is the assumption of linear changes in achievement over time. The model assumes such a linear relation, but the changes in achievement from 2011 to 2019 are not always linear. For instance, in Sweden, achievements inclined much more between 2011 and 2015 (15 points) than between 2015 and 2019 (2 points). The regression coefficient between time and achievement in our models reflects a *mean* growth from 2011 to 2015. This prevents us from providing detailed information about the differences in growth between each cycle. Our findings instead reflect mean changes across the whole period from 2011 to 2019.

The OTL measures have changed over time, as has the test. While this is reasonable, it makes it harder to connect to changes in the countries' curricula. However, in this study, we have only used those content coverage items that have been repeated over all three cycles, which is the major part. Not including the new content coverage items may have limited the coverage of each sub-topic. Further research is warranted to investigate this possibility.

The large number of missing responses from the teachers in Denmark, Norway, and Sweden, to the questions about content coverage in science, prevented a proper analysis of OTL in science. The countries may want to encourage and inform teachers of the importance of the OTL measures for research and policy (even if there are many items for the teachers to respond to). Also, these rather demanding questions regarding content coverage in science to the teacher are currently located late in the teacher questionnaire, relocation of these should be considered for a more optimal response rate.

6.6.2   Concluding Remarks, Contributions, and Implications

The main aim of the present study was to examine whether changes in OTL with respect to content coverage were related to changes in achievement. Our findings confirmed this, and this may imply that changing the implemented curriculum may have consequences for students' competence. All countries have a national curriculum that all teachers should follow. Given the within-country variation in achievement, there is a need to investigate to what degree this variation may be due to a lack of content coverage of the assessed curriculum in the schools. Lack of, or variation in content coverage in schools implies an unequal opportunity for students to learn, and unequal opportunities to be fairly assessed. The topic of unequal opportunities for students to learn is further examined in Chap. 8.

The findings from this study contribute to teacher education, stakeholders in education and educational policy, and curriculum makers, as content coverage considers the alignment between what is taught and what is assessed and the alignment with the curriculum. The study further contributes to research, more specifically to OTL research. The content coverage variables in TIMSS are far from optimal from a measurement point of view. We agree with Scheerens's (2016) conclusion that there is a need to address the methodological challenges attached to the OTL indicators. A closer examination of the validity of the OTL measures in TIMSS, and actions and methods to improve the measurement properties of the indicators are urgently needed. Actions include considerations of the questionnaire design, the location of the OTL items, the phrasing and the translations of the items, and some validation studies to ensure that the questions and the response scales are working as intended and interpreted in the same way by the responding teachers.

The study also contributes to research within assessment, measurement, and psychometrics. The methodological approach should interest researchers who wish to examine relations between changes in predictors and outcome variables, such as achievement in other ILSAs and/or other countries.

Appendices

Appendix 1 How Changes in OTL in Number, Geometry, and Data Are Related to Changes in Mathematics Achievement Over Time

Mediating variable	The effect of the mediating variable on math achievement				The effect of time on predictor				The effect of time on math achievement (direct effect)				Indirect effect			
	SWE	NOR	FIN	DEN	SWE	NOR	FIN	DEN	SWE	NOR	FIN	DEN	SWE	NOR	FIN	DEN
OTL number topics	0.1	0.4^{**}	0.2^{**}	0.3^{**}	5.9^{**}	8.0^{**}	3.3^{**}	4.3^{**}	10.5^{**}	19.3^{**}	-6.6^{**}	-7.3^{**}	0.8	3.2^{**}	0.6^{**}	1.2^{**}
OTL geometry topics	0.1	0.3^{**}	0.1^{**}	0.3^{**}	9.5^{**}	-5.6^{**}	7.3^{**}	0.0	10.7^{**}	24.0^{**}	-6.9^{**}	-6.1^{**}	0.6	-1.5^{**}	0.8^{**}	0.0
OTL data topics	0.0	0.1^{**}	0.1^{*}	0.1^{**}	-4.2^{*}	10.2^{**}	-11.3^{**}	2.7	11.3^{**}	21.3^{**}	-5.4^{**}	-6.5^{**}	0.0	1.3^{**}	-0.6^{*}	0.3

Appendix 2 Percentages of Teachers/classrooms with no Responses to the Content Coverage Items in Science

	Life science			Physical science			Earth science		
	2011	2015	2019	2011	2015	2019	2011	2015	2019
Denmark	32.7	16.9	20.0	32.9	17.3	21.1	32.0	17.4	19.8
Finland	4.8	4.7	2.6	5.2	4.7	3.1	4.8	5.1	3.0
Norway	7.2	23.1	36.0	8.0	23.4	36.0	7.2	23.8	36.0
Sweden	28.4	7.1	15.1	30.0	6.2	14.3	31.2	4.6	15.7

Note The information about the missingness is selected from the almanacs provided by TIMSS & PIRLS International Study Centre, Boston College

Appendix 3 Indirect Effects and the Total Indirect Effect for the Model Where All Three OTL Variables in Science Were Included Simultaneously in One Model

	SWE	NOR	FIN	DEN
Indirect effect. OTL in Life science	0.0	0.4	0.0	0.0
Indirect effect. OTL in Physical science	0.1	0.3	0.0	0.1
Indirect effect. OTL in Earth science	0.3	− 0.6	0.0	0.2
Total indirect effect. All OTL variables included	0.4	0.2	0.0	0.1

Note None of the indirect effects are significant

References

Caroll, J. B. (1963). A model of school learning. *Teachers College Record: The Voice of Scholarship in Education, 64*(8), 722–733. https://doi.org/10.1177/016146816306400801

Dahllöf, U. (1970). Curriculum process analysis and comparative evaluation of school systems. *Paedagogica Europaea*, 21–36. https://doi.org/10.2307/1502497

Fishbein, B., Foy, P., & Yin, L. (2021). *TIMSS 2019 user guide for the international database* (2nd ed.). Retrieved from Boston College, TIMSS & PIRLS International Study Center website: https://timssandpirls.bc.edu/timss2019/international-database/

Gustafsson, J.-E. (2010). Longitudinal designs. In B. P. Creemers, L. Kyriakides, & P. Sammons (Eds.), *Methodological advances in educational effectiveness research* (pp. 77–101). Routledge.

Gustafsson, J.-E., & Nilsen, T. (2022). Methods of causal analysis with ILSA data. In T. Nilsen, A. Stancel-Piątak, & J.-E. Gustafsson (Eds.), *International handbook of comparative large-scale studies in education*. Springer International Handbooks of Education. https://doi.org/10.1007/978-3-030-38298-8_56-1

Husen, T. (1967). *International study of achievement in mathematics: A comparison of twelve countries*. Wiley.

Husén, T., & Dahllöf, U. (1965). An empirical approach to the problem of curriculum content. *International Review of Education, 11*(1), 51–76. https://doi.org/10.1007/BF01421682

McDonnell, L. M. (1995). Opportunity to learn as a research concept and a policy instrument. *Educational Evaluation and Policy Analysis, 17*(3), 305–322. https://doi.org/10.3102/016237 3701700330

Mullis, I. V. S., Martin, M. O., Foy, P., Kelly, D. L., & Fishbein, B. (2020). *TIMSS 2019 international results in mathematics and science*. Retrieved from Boston College, TIMSS & PIRLS International Study Center website: https://timssandpirls.bc.edu/timss2019/international-res ults/

Murnane, R. J., & Willett, J. B. (2010). *Methods matter: Improving causal inference in educational and social science research*. Oxford University Press.

Muthen, L. K., & Muthen, B. O. (2017). MPlus (Version 8.1) [Computer software]. Muthen & Muthen. https://www.statmodel.com/index.shtml

Olsen, R. V., & Bjørnsson, J. K. (2018). Fødselsmåned og skoleprestasjoner [Month of birth and school performance]. In *Tjue år med TIMSS og PISA i Norge: Trender og nye analyser* [Twenty years with TIMSS and PISA in Norway: Trends and new analyses] (pp. 76–93). Universitetsforlaget. https://doi.org/10.18261/9788215030067-2018-05

Pelgrum, W. J. (1989) *Educational assessment. Monitoring, evaluation and the curriculum*. Academisch Boeken Centrum.

Scheerens, J. (Ed.). (2016). *Opportunity to learn, curriculum alignment and test preparation: A research review*. Springer. https://doi.org/10.1007/978-3-319-43110-9

Scherer, R. (2022). Analyzing international large-scale assessment data with a hierarchical approach. In T. Nilsen, A. Stancel-Piątak, J.E. Gustafsson (Eds.), *International handbook of comparative large-scale studies in education: Perspectives, methods and findings* (pp. 1–55). Springer International Publishing. https://doi.org/10.1007/978-3-030-38298-8_59-1

Schmidt, W. B., Cogan, L. S., Houang, R. T., & McKnight, C. C. (2011). Content coverage across countries/states: A persisting challenge for US educational policy. *American Journal of Education, 117*(3), 399–427. https://doi.org/10.1086/659213

Schmidt, W. H., Cogan, L. S., Houang, R. T., & McKnight, C. (2009). *Equality of educational opportunity: A myth or reality in US schooling*. Education Policy Center at Michigan State University.

Schmidt, W. H., Cogan, L. S., & McKnight, C. C. (2010). Equality of educational opportunity: A myth or reality in US schooling. *American Educator 2010–2011*.

Monica Rosén is a Professor of Education at the University of Gothenburg. She has been involved in research on international large-scale assessment since the 1990s. Her main areas of research concern educational results at individual-, group-, and system levels, and factors associated with differences and changes in educational outcomes. She also shares a strong interest in the methodological issues following these research interests. She is currently the head of three national research schools which focuses on educational assessment and quantitative research methods.

Trude Nilsen is a research professor at the University of Oslo. She is a leader of Strand 1 at CREATE - Centre for Research on Equality in Education, a leader of the research group LEA, and of funded research projects. She has been engaged as an international external expert for IEA's TIMSS and for OECD's TALIS studies. Her research focuses on teaching quality, educational equality, school climate, and applied methodology including causal inferences.

Chapter 7
Changes in Teacher Practices Related to Changes in Student Achievement

Trude Nilsen⊙ and Jan-Eric Gustafsson⊙

7.1 Introduction

One of the most useful and reliable results from international large-scale assessment (ILSA) is the changes in student achievement across time within countries. While comparisons across countries can sometimes be problematic due to cultural differences and variations between the educational systems, comparisons within countries over time can be very useful for policy and practice. Knowledge of *why* achievement changes over time can be as informative and useful as making cross-national comparisons at any point in time. It is, however, difficult to provide evidence of the reasons behind changes in achievement. Most ILSAs are cross-sectional, and hence causal inferences based on these data are limited. However, there are so-called causal methodologies that provide more robust and reliable results and that increase the plausibility for causal inferences (Gustafsson & Nilsen, 2022; Nilsen et al., 2022). The present study utilizes such an approach to investigate plausible reasons behind the changes in science and mathematics achievement in the Nordic countries over the last three cycles (2011, 2015, and 2019) of the Trends in Mathematics and Science Study (TIMSS).

Several factors may have caused changes in student achievement. From previous research, we know that teachers are the most proximal to students, and key to student

T. Nilsen (✉)
Department of Teacher Education and School Research, CREATE—Centre for Research on Equality in Education (project number 331640), University of Oslo, Postboks 1099, Blindern, 0317 Oslo, Norway
e-mail: trude.nilsen@ils.uio.no

J.-E. Gustafsson
Department of Education and Special Education, University of Gothenburg, P.O. Box 300, SE 40530 Gothenburg, Sweden
e-mail: Jan-Eric.Gustafsson@ped.gu.se

N. Teig et al. (eds.), *Effective and Equitable Teacher Practice in Mathematics and Science Education*, IEA Research for Education 14,
https://doi.org/10.1007/978-3-031-49580-9_7

learning outcomes (Baumert et al., 2010; Charalambous & Praetorius, 2020; Darling-Hammond, 2000; Nilsen & Gustafsson, 2016). Especially, teacher practices are known to shape student learning, well-being, and motivation (Fauth et al., 2014; Senden et al., 2022). Teacher practices are part of a broad concept that may include teaching quality and homework practices. Teaching quality is defined as the aspects of teaching, that previous research has found, that promote learning, and refers to *what teachers do in the classroom* (Praetorius et al., 2018). Homework practices may refer to how often and how long (number of minutes) the homework assigned to students is, the type of homework assigned, and what the teachers do with the homework, such as discussing the homework in class or correcting the homework (Fernández-Alonso & Muñiz, 2021). Previous research on the effect of homework practices on student achievement is mixed; some research finds that homework promotes learning and others don't (Fan et al., 2017; Gustafsson, 2013; Trautwein, 2007). This can often depend on the conceptualizations and methodologies used in the research.

The overarching aim of the present study is to investigate whether changes in teacher practices are related to changes in student science and mathematics achievement from 2011 to 2019 using TIMSS data.

7.2 Theoretical Framework

In this section we define the concepts included in this study based on previous research and describe what previous research has found in terms of relations between teacher practices and student learning outcomes.

This study defines teacher practices to encompass teaching quality, homework, and assessment practices, as detailed in Chap. 2. However, assessment practices are only included in one of the three cycles and are hence excluded from this chapter.

7.2.1 Teaching Quality

There are several frameworks describing teaching quality, and the concept itself has many names, such as instructional quality, teaching quality, and teacher practices. This is problematic and has hence been a topic of debate recently (Charalambous & Praetorius, 2020; Senden et al., 2022). The frameworks describing teaching quality are different depending on the type of data. For instance, frameworks describing teaching quality in studies using video observations of classrooms may differ from those using questionnaires (Senden et al., 2022). However, the key aspects of teaching quality are very similar. The present study uses the framework of the three basic dimensions (TBD) by Klieme and colleagues (Klieme et al., 2009; Praetorius et al., 2018), as this has been tested and validated in several studies in Europe (Senden et al., 2022, 2023).

Teaching quality refers to the aspects of teaching, more specifically the teachers' behavior in the classroom, that in previous studies have been found to promote learning outcomes (Baumert et al., 2010; Fauth et al., 2014). The TBD framework includes three main aspects of teaching. *Classroom management*, referring to maintaining order and discipline in class and time on task, is the strongest predictor of achievement (Marder et al., 2023). TIMSS included this measure in the student questionnaire for the first time in 2019. *Teacher support and clarity* refers to both social-emotional support and support for learning, as well as clear and understandable teaching (Praetorius et al., 2018). This second aspect of teaching has been found to be a stronger predictor of student motivation than student achievement (Fauth et al., 2014; Nilsen et al., 2018). The third aspect of teaching quality is *cognitive activation*, which refers to challenging students to go beyond what they have learned, to promote reflection, reasoning, and critical thinking. In mathematics, problem-solving is an example of cognitive activation. In science, inquiry is a typical approach that would challenge students cognitively, for instance, by making students plan and interpret findings from an experiment (Teig et al., 2018, 2021). This aspect has been found to promote student learning, although findings are mixed as it is hard to measure (Baumert et al., 2010; Charalambous & Praetorius, 2020).

The TBD framework is a generic framework that works well across subject domains, and all three aspects have been found to be relevant in both mathematics and science (Fauth et al., 2014; Praetorius et al., 2018). However, cognitive activation is more content specific than the other two (Praetorius et al., 2018).

7.2.2 Interaction Between the Teaching and the Students—Limitations to Teaching Quality

The teaching that goes on in the classroom is not a one-way street; the quality of teaching depends not just on the teacher, but also on the class. The classroom composition matters. For instance, a classroom dominated by students who lack prior knowledge, who do not speak the language well, or who are hungry or sleepy, would limit high quality teaching (Kaarstein & Nilsen, 2021). Even if the teacher is competent and usually provides high quality teaching, such limitations would be challenging. Hence, considering such limitations when examining teaching quality is necessary.

7.2.3 Homework

Findings from research on the effects of homework are mixed (Fernández-Alonso & Muñiz, 2021; Gustafsson, 2013; Trautwein, 2007). This may be explained by several factors, such as the type of homework. For instance, in mathematics, teachers may provide homework that includes repetitions of tasks the students have done at school, or they may provide cognitively challenging tasks that would require reasoning or problem-solving. Further, the teacher may provide homework often, but could assign tasks that may be done quickly by students, or they could assign homework more seldom but that could take the students a long time to complete. Low-achieving students, and/or students from homes of low socioeconomic status (SES), usually need longer time to complete a homework task than high-SES students (Gustafsson, 2013). The findings may also depend on the operationalization of homework, the sampling design, the level of analyses, and the methods of analyses. For instance, using data with a cross-sectional design, operationalizing homework by the number of minutes students spend on homework, and analyzing the effect of homework on achievement at the student level, may produce negative regression coefficients. Low achieving students often use more time on homework, and hence, a longitudinal study would be needed to control prior achievement to enable reliable results. However, with data aggregated to the class-level, the impact of characteristics of individual students is reduced, and the influence of teacher-assigned homework may be expected to be more pronounced (Gustafsson, 2013).

7.3 Research Questions

The overarching aim of the present study is to investigate whether *changes in teacher practices are related to changes in science and mathematics achievement from 2011 to 2019*. Seeing how Norway changed the target grade in 2015 from grade four to grade five, the aim for Norway differs somewhat, and we investigate whether *changes in teacher practices are related to changes in science and mathematics achievement over time (from 2011 to 2015) and/or changes in achievement from grade four to five*. We address this aim more specifically through the following research questions:

1. *How have teacher practices changed from 2011 to 2019?*
2. *Are changes in teachers' practices related to changes in achievement? In Sweden, Denmark, and Finland the changes in achievement refer to changes from 2011 to 2019, while in Norway, the changes in achievement refer to changes from 2011 to 2019, and/or changes from grade four to five.*

Achievements have increased in some countries during this time span and decreased for others. For the countries where achievements have increased, this could be related to an increase in one or several of the teachers' practices. In other words, the increased achievements may be explained by better teaching practices over time. For this to happen, then the increased achievement must be *related* to increased teacher practices. The same is true for decreased achievements and teacher practices. If teacher practices have not changed, or if they are not related to achievement, they cannot explain any changes in achievement.

7.4 Methodology

7.4.1 Data and Sample

The data used in the present study is achievement and questionnaire data from representative samples of students from Sweden, Norway, Finland, and Denmark as well as their teachers. These data are retrieved from three cycles of TIMSS: 2011, 2015, and 2019. In Sweden, Finland, and Denmark, the students who participated were in grade four. In Norway, the target grade changed in 2015 from fourth to fifth grade, so that the present sample includes Norwegian fourth graders in 2011, and fifth graders in 2015 to 2019. The reason why the target grade changed from grade four to five in Norway, was because the Norwegian students in grade four were one year younger than students in grade four in the other Nordic countries. This is because Norwegian students start school the year they turn six years old, while students from the other Nordic countries start school the year they turn seven years old. Hence, to compare Norwegian results from TIMSS with that of other Nordic countries, Norway changed the target grade to grade five. This means that findings from Norway need a different interpretation. Part of the explanation for the increased achievement in Norway from 2011 to 2019, is that students were older and had one more year of schooling (Olsen & Björnsson, 2018). For instance, if findings show that teacher practices explain part of the changes in achievement, it means that teacher practices may explain changes in achievement across time and/or across grades. For descriptions of TIMSS data, including sampling, plausible values, and weights, please see Chap. 3.

7.4.2 Measures

Some of the measures in the TIMSS questionnaires have changed over time, especially teaching quality. To be able to compare measures across time, only items that have stayed the same over time are included.

Teacher practices

Several variables were used to measure the different aspects of teaching quality. Teacher support and clarity were measured by students' responses to their perceptions of their teachers (e.g., I know what my teacher expects me to do, my teacher is easy to understand). Cognitive activation in mathematics was measured by teachers' responses to how often they ask students to do various tasks (e.g., listen to me explain how to solve problems, work problems together in the whole class with direct guidance from me). The same question was asked for science (e.g., observe natural phenomena such as the weather or a plant growing and describe what you see, design or plan experiments or investigations). Classroom management was not introduced before 2019 and was thus excluded from the study.

Limitations to teaching were measured by teachers' responses to the extent their teaching was limited by various challenges (e.g., students lacking prerequisite knowledge or skills, students suffering from lack of basic nutrition, students suffering from not enough sleep).

There were two measures for homework. The first is related to how teachers use the homework, and this was measured by teachers' responses to how often they do the following with the homework: correct assignments and give feedback to students, discuss the homework in class, monitor whether or not the homework was completed. In addition, the teachers were asked how often (frequency) and how many minutes of homework they assign. To create an estimate of time spent on mathematics homework, an algorithm was created which multiplied the categories in the teacher questionnaire of frequency and time into an estimate of minutes spent on homework each week by each student (see Gustafsson, 2013, for a description of this procedure). In a parallel fashion an algorithm for estimating time spent on science homework was created.

A relatively large number of teachers reported that homework was never assigned, and particularly so for science. The proportion of teachers reporting any homework in mathematics was for Denmark 72%, Finland 86%, Norway 68%, and Sweden 65%. For science, the proportion of teachers reporting any homework was for Denmark less than 10%, Finland 88%, Norway 34%, and Sweden 33%. Thus, for science only, Finland assigns homework to such an extent that meaningful analyses can be conducted from the data. For mathematics, results must also be cautiously interpreted, given that only about two thirds of the students in Denmark, Norway and Sweden were assigned any homework.

7.4.3 The Analytical Approach

The method of analyses in this study resembles that of longitudinal growth models (Murnane & Willett, 2010) but adjusted to examine trend analyses.

Causality and causal language. The method of analyses implemented in this study utilizes merged data from three cycles. The method is far more robust than analyzing one cycle of an ILSA with cross-sectional data, or by analyzing data from each cycle separately, and then comparing results across time. The present method enhances the plausibility of causal inferences. That being said, the design is not truly longitudinal, as the same students are not followed over time. Yet, taking advantage of the representative samples and the trend-design of TIMSS—where achievements are scaled to enable comparisons across time—the reliability and validity of inferences is enhanced.

Still, this is not a randomized controlled trial, and hence, the inferences are not strictly causal. This means there is a need to examine the findings in light of previous research, and to consider omitted variables that may introduce bias. This further implies that causal language should not be used. However, this book is aimed at

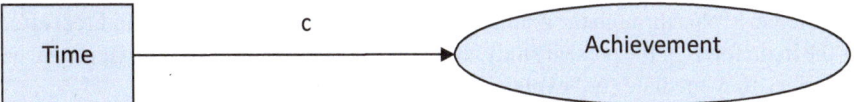

Fig. 7.1 The effect of time on achievement using the null model. *Note* The regression coefficient is c

education policy, stakeholders, and researchers. To clarify advanced methodologies, terms like effect are used to describe the relationship between predictors and outcomes. Nevertheless, both in this chapter and the conclusions, it is emphasized that the inferences are not based on causal relationships.

The analytical approach. The approach in the present chapter is that of a structural equation model (SEM) with mediation, and we use the software Mplus for the analyses (Muthén & Muthén, 1998–2017). A dummy variable called *Time* is coded 0 for 2011, 1 for 2015, and 2 for 2019. In a null model we first examine changes in achievement over time as described in Fig. 7.1.

The regression coefficient c reflects how achievement has changed over time and describes the slope of the effect of time on achievement.

The idea is to examine whether there are other factors that may "explain" changes in achievement, besides the passing of time. In this case, we wish to examine whether teacher practices may explain changes in achievement. We do this by examining whether teacher practices may *mediate* the effect of time on achievement as illustrated in Fig. 7.2.

If an aspect of teacher practices (e.g., teaching quality) mediates the effect of time on achievement, this means that changes in teacher practices are related to changes in achievement, and it may indicate that changes in teacher practices explain changes in achievement over time.

If teacher practices have improved over time, the regression coefficient a would be positive, and if it is positively related to achievement, it means the regression coefficient b is positive. If both coefficients a and b are statistically significant, then the conditions for mediation are present. Whether or not the mediation is significant,

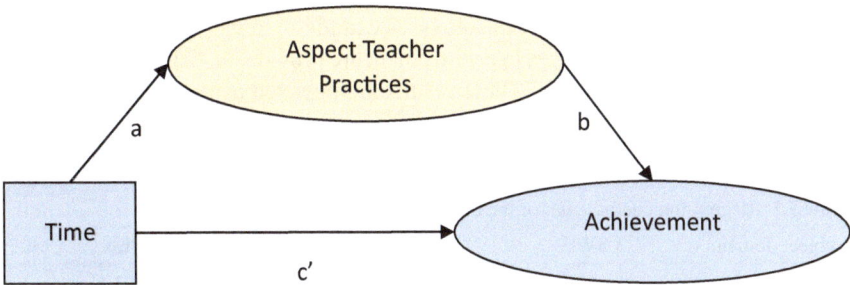

Fig. 7.2 Hypothesized mediation model in which aspects of teachers' practices mediate the effect of time on achievement

is tested in Mplus through the command *Model indirect*. This so-called indirect effect is the main focus of the present study, as it tells us whether, and to what extent, teacher practices may mediate (or "explain") changes in achievement.

Seeing how the aim is to explain changes in *students'* achievement over time, and not differences in achievement between classes, the analyses are done at the student level. However, to take into account the hierarchical design of the data where students are nested within classes that are nested within schools, and to avoid under-estimation of standard errors, we use the option in Mplus called "TYPE = COMPLEX" where the data are clustered at the class-level. This way, the analyses take the hierarchical clustering of students into account. With regards to socioeconomic status (SES), it is a fixed effect, as it does change over time and is hence redundant in the analyses.

7.5   Results

In regard to measures of reliability and validity, the model fit was good (i.e., satisfying the cut-off points: the root of mean square error approximation [RMSEA] < 0.05, comparative fit index [CFI] > 0.95), and the factor loadings of the latent variable were between 0.51 and 0.92.

The null model. Before introducing the mediator and predictor of teacher practices, we examined the effect of time on achievement. The results reflect the slope of the relation between time and achievement. If the regression coefficient is positive, it means that achievement increased over time, and a large positive coefficient would reflect a large increase. Table 7.1 shows the results of the null model for science and mathematics.

The findings show an increase in science and mathematics achievement from 2011 to 2019 for Sweden and Norway. For Norway, part of the large increase was because the target grade changed from grade four to five in 2015 (Olsen & Björnsson, 2018). For Finland and Denmark, the achievements decreased in both subject domains. These findings are in line with the international TIMSS report (Mullis et al., 2020). However, Table 7.1 reflects regression coefficients for the slope of changes over two cycles. The actual changes in achievement are available in Chap. 1.

The full model. The results of the analyses examining relations between changes in teacher practices and changes in achievement are provided in Table 7.2 for science and Table 7.3 for mathematics. These results are depicted as text and symbols, the numbers are provided in Appendices 1 and 2.

Table 7.1 Regression coefficients for the effect of time on achievement

Subject domain	SWE	NOR	FIN	DEN
Science	5.3^*	22.1^*	-6.3^*	-3.1^*
Mathematics	11.3^*	22.6^*	-6.1^*	-6.1^*

Table 7.2 How changes in teacher practices are related to changes in science achievement over time

	What is the effect of teacher practices on science achievement?				How have teacher practices changed over time? (have they decreased ↓ or increased ↑)				How are changes in teacher practices related to changes in achievement (indirect effect)?			
	SWE	NOR	FIN	DEN	SWE	NOR	FIN	DEN	SWE	NOR	FIN	DEN
Clarity of instruction	NS	Small, positive effect	Medium, positive effect	NS	↓	↓	NS	↓	NS	↓ by 1 point	NS	NS
Cognitive activation (inquiry)	Large, positive effect	Large, positive effect	Small, positive effect	NS	↓	↑	NS	NS	↓ by 1 point	↑ by 2 points	NS	NS
Limitations to teaching	Medium, positive effect	Medium, positive effect	Large, positive effect	Medium, positive effect	↓	↓	↓	↓	↓ by 7 points	↓ by 3 points	↓ by 9 points	↓ by 7 points
Use of homework	NS	NS	NS	NS	NS	NS	↓	NS	NS	NS	NS	NS
Time spent on homework	Small, negative effect	NS	Small, positive effect	(grey)	↓	NS	↓	(grey)	NS	NS	NS	(grey)

Note If the regression coefficient is below 10, this is denoted a small effect, 10 to 20 = medium, over 20 = large. The symbols ↓ denotes a decrease, ↑ denotes an increase. The grey cells denote models that did not converge

Table 7.3 How changes in teacher practices are related to changes in mathematics achievement over time

	What are the effects of teacher practices on mathematics achievement?				How have teacher practices changed over time? (Have they decreased ↓ or increased ↑)				How are changes in teacher practices related to changes in achievement (indirect effect)?			
	SWE	NOR	FIN	DEN	SWE	NOR	FIN	DEN	SWE	NOR	FIN	DEN
Clarity of instruction	Medium, positive effect	Large, positive effect	Large, positive effect	Large, positive effect	→	→	←	→	↓ by 1 point	↓ by 3 points	↑ by 1 point	↓ by 1 point
Cognitive activation	NS	NS	NS	NS	←	→	NS	NS	NS	NS	NS	NS
Limitations to teaching	Medium, positive effect	Medium, positive effect	Large, positive effect	Medium, positive effect	→	→	→	→	↓ by 6 points	↓ by 3 points	↓ by 9 points	↓ by 6 points
Use of homework	NS	NS	NS	NS	NS	NS	→	NS	NS	NS	NS	NS
Time spent on homework	Small, negative effect	NS	NS	Small, positive effect	→	NS	→	→	↑ by 1 point	NS	NS	NS

Note If the regression coefficient is below 10, this is denoted a small effect, 10 to 20 = medium, over 20 = large. The symbols ↓ denotes a decrease, ↑ denotes an increase

Science. In Table 7.2, the results are provided for the different teacher practices. In the first two rows, we explain in detail the first two aspects of teacher practice, clarity of instruction and cognitive activation, and thereafter we only comment on noteworthy findings. The first aspect of teacher practice, clarity of instruction (reported by students), has a positive and statistically significant relation to science achievement in Norway (small effect) and Finland (medium effect), while findings for Denmark and Sweden were not significant. Students' perception of clarity of instruction decreased from 2011 to 2019 in all countries, except for Finland. This means that students perceived the instruction as less clear in 2019 than in 2011. To address the main focus of the present study, i.e., whether changes in teacher practices are related to changes in achievement, changes in the clarity of instruction were only significantly related to changes in achievement in Norway. In Norway, the slope of the relation between time and achievement was about 22 points, meaning that achievement increased by 22 points from 2011 to 2019 (where parts of this large increase in achievement is explained by the change of target grade from grade four to grade five). The indirect effect was negative and about one point. This may indicate that Norway could have increased their achievement by 23 points (rather than 22 points) had it not been for the decrease in clarity of instruction. This relation, between changes in clarity of instruction and changes in achievement, could indicate that decreased clarity of instruction may explain changes in achievement over time, and/or changes in achievement from grade four to grade five.

Cognitive activation in science, more specifically inquiry, as reported by teachers, had a significant and positive relation to science achievement in all countries, except for Denmark. However, it only changed significantly over time in Sweden and Norway. It decreased in Sweden and increased in Norway, meaning that teachers reported providing less inquiry-based teaching in 2019 than in 2011 in Sweden. In Norway, it could mean that teachers provided more inquiry-based teaching in 2019 than in 2011, and/or that teachers provided more inquiry-based teaching in grade five than in grade four. In both these countries these changes were significantly related to changes in achievement. The relation between time and achievement in Sweden was about five points, meaning that achievement increased by five points. The indirect effect was negative, and about one point. This may indicate that Sweden could have increased their science achievement by six points (rather than five points), had it not been for the decrease of cognitive activation. In Norway, the indirect effect was positive and about two points. This may indicate that Norway would only have had an increase of 20 points in their science achievement (rather than 22), had it not been for the increase in cognitive activation. This relation between changes in inquiry-based teaching and changes in achievement could indicate that the increase in cognitive activation may explain changes in achievement over time, and/or changes in achievement from grade four to grade five.

Table 7.2 further provides results on limitations to teaching, how homework is used in class, and the time and frequency of homework. For these, there are two noteworthy findings. First, limitations to teaching worsened in all countries, meaning that the teachers' reported that the following aspects limited instruction more in 2019 than they did in 2011: more students lacking sleep, more students being hungry in school, less students with prior knowledge, and more students disturbing and being less interested in the instruction. This negative change is related to changes in achievement over time. In Sweden, limitations to teaching decreased achievement by seven points, in Norway by three points, in Finland by nine points, and in Denmark by seven points. This means that, for instance, Finland could have turned their decreased achievement, to increased achievement. For Norway, changes in limitations to teaching could explain either changes in achievement over time or over grades, or both. Second, the data show that very little science homework was assigned in grade four. Except for Finland, most students were not assigned any homework at all and for those who do get homework the assignments only require between three and 11 min of work each week. Even for Finland, where 88% of the students were assigned homework, the weekly average spent on science homework was eight minutes, or less than two minutes per day. Such small doses of science homework can hardly be expected to yield any significant effect, because of low statistical power. As expected, there was no significant indirect effect of homework on science achievement. The means of homework for the Nordic countries, and how these change over time, is provided in Fig. 5.3 in Chap. 5.

Mathematics. Table 7.3 provides results for relations between changes in teacher practices and changes in mathematics achievement. As for science, in the first two rows we also explain in detail the first two aspects of teacher practice, clarity of instruction and cognitive activation, and thereafter we only comment on noteworthy findings for mathematics. The first aspect of teacher practice, clarity of instruction (reported by students) has a positive and significant relation to mathematics achievement in all countries. Students' perception of clarity of instruction decreased from 2011 to 2019 in all countries, except for Finland. For Norway, clarity of instruction could have changed over time or over grades, or both. Changes in clarity of instruction were significantly related to changes in achievement in all countries. In Sweden, Norway, and Denmark the indirect effects were negative, while in Finland it was positive. This means that Norway and Sweden's inclining achievements may have increased even further, and Denmark's achievement could have declined less, if clarity of instruction had increased rather than decreased for these three countries. In Finland, on the other hand, clarity of instruction increased, and this may have prevented a further decrease of their achievements.

Cognitive activation in mathematics was not significantly related to mathematics achievements in any of the countries. This is probably due to the low validity of the construct that only includes two items. However, it increased significantly over time in Sweden. In Norway it decreased, either over time or over grades, or both. There were no significant indirect effects, meaning that changes in cognitive activation are not related to changes in achievement.

Table 7.3 further provides results on limitations to teaching, how homework is used in class, and the time and frequency of homework. As for science, there are two noteworthy findings. First, limitations to teaching worsened over time in all countries (in Norway it decreased over time and/or from grade four to five). This means that the teachers' reported that the following aspects limited instruction more in 2019 than they did in 2011: more students lacking sleep, more students being hungry in school, less students with prior knowledge, and more students disturbing and being less interested in the instruction. This negative change is related to changes in achievement over time (and/or change in grades in Norway). In Sweden, limitations to teaching decreased changes in achievement by six points, in Norway, by three points, in Finland by nine points, and in Denmark by six points. This means that, for instance, Finland could have turned their decreased achievement, to increased achievement. Again, changes in limitation to teaching in Norway, could be related to changes in achievement over time and/or grade.

The amount of homework assigned in mathematics was somewhat larger than in science, and particularly so for Norway and Denmark (24 and 14 min, respectively). The weekly mean was 13 and nine minutes for Finland and Sweden, respectively. No significant indirect effect of homework was found for mathematics, but for Sweden an effect of 1.4 ($t = 1.92$) was found. However, until this borderline significance has been replicated it should be regarded as a chance effect, given the large number of statistical tests conducted.

7.6 Discussion

7.6.1 Interpretation of Findings and Discussion in Light of Previous Research

Relations to achievement. In both mathematics and science, there is a strong pattern of positive and significant relations between teacher practices and student achievement, except for homework where there were no/few significant findings. The positive, significant relations between clarity of instruction and student achievement in both subject domains are in line with previous research (Fauth et al., 2014; Nilsen et al., 2018). The relation between cognitive activation and achievement was only positive and significant in science but was insignificant in all countries in mathematics. In science, cognitive activation was measured in terms of inquiry, and the positive results have been identified in previous research (Teig et al., 2018). The measure of inquiry has remained the same across all cycles and includes five items. The measure of cognitive activation in mathematics, however, has changed substantially over time; only two items were equal over time (i.e., how often the teacher asks the students to: "listen to me explain how to solve problems," and "work problems (individually or with peers) with my guidance"). The validity of this measure is hence very low, and we believe this may explain the insignificant findings.

Besides the aspects of teaching quality, there was also a strong pattern of significant and positive relations across all countries and in both subject domains between limitations to teaching and student achievement. This means that students with teachers who perceived less limitations to teaching, perform better. This finding is supported by previous research (Kaarstein & Nilsen, 2021; Vik et al., 2022a).

Changes in teacher practices over time. In regard to changes in clarity of instruction, there is a clear pattern across both subject domains: students perceive their teachers to provide less clear instruction in 2019 than in 2011. The exception is Finland: in mathematics, Finnish students perceive their teachers to provide clearer instruction in 2019 than in 2011, while there are no significant changes in science.

However, there is no clear pattern of findings for cognitive activation, except that in both subjects there are no significant changes in Denmark or Finland. In science, teachers reported implementing inquiry practices more often in 2019 than in 2011 (and/or in grade five than in grade four) in Norway, and less often in Sweden. In mathematics, teachers reported implementing cognitive activation less often in 2019 than in 2011 in Norway (and/or in grade five than in grade four) and more often in Sweden. These opposite results and lack of patterns across mathematics and science are not surprising, due to the aforementioned low validity of the measure of cognitive activation in mathematics.

Findings on limitations to teaching show a clear pattern; a negative change in all countries. This means that teachers reported more limitations to teaching in 2019 than in 2011 (and/or more limitations to teaching in grade five than grade 4 in Norway). The samples in TIMSS are representative, so this could imply that the composition and habits of the populations of students have changed over time. To a large extent, this is backed by previous research; students' habits in terms of gaming and social media have changed over time, and this could cause lack of sleep (Vik et al., 2022b). Moreover, there was a larger number of minority students who did not speak the language of the test as well as majority students did in 2019 as compared to in 2011 (Mullis et al., 2020). However, more research is needed to investigate how the student population has changed over time, and the consequences of this change.

Changes in teacher practices related to changes in student achievement. Clarity of instruction decreased over the time period for all countries except for Finland. However, if the change is not related to achievement, and if there is no significant indirect effect, this change cannot explain changes in achievement. For science, the indirect effect was only significant for Norway, and the effect was small and negative (one point). Norway's increased achievement of 22 points (due to changes in time and/or grade), would have been 23 points had it not been for the decrease in clarity of instruction. For mathematics, on the other hand, the changes in clarity of instruction were related to changes in achievement in all countries. This could indicate that clarity of instruction is more important for student achievement in mathematics than in science. This is indeed confirmed by our findings through the stronger relations between clarity of instruction and achievement in mathematics than in science. While clarity of instruction is also important in science (Nilsen et al., 2018), it could be that other aspects of teaching quality are also important in science, such as inquiry-based teaching. Our findings confirm that the aspect of cognitive activation (inquiry), is positively associated with student science achievement, as has been found in previous research (e.g. Teig et al., 2018). This aspect of teaching quality in science only changed over time in Sweden and over time and/or grade in Norway. In Sweden, the increase in achievement would have been one point, had it not been for the decrease in inquiry. In Norway, the increase in inquiry explained two of the 22 points of their increased achievement over time and/or grade.

The analyses of effects of time spent on homework in science and mathematics did not yield any significant effects. However, this lack of findings must be interpreted cautiously, given that only limited amounts of homework were assigned to students, and particularly so for science. The fact that only a few students were actually assigned homework caused the statistical power to be limited, in spite of the relatively large samples of students involved in the study.

The most striking results of our analyses are those of limitations to teaching. As pointed out earlier, less limitations to teaching were positively associated with achievement, but in all countries, the teachers perceived more limitations to teaching in 2019 than in 2011. If it had not been for this negative development of limitations to teaching, Finland and Denmark could have turned their negative developments of achievements to the positive, and in Norway and Sweden, their increased achievements would have been increased even more.

In conclusion, our findings indicate that while teaching quality is important for student achievement, there have been no major changes in teaching quality, and these small changes only explain minor parts of changes in achievement. However, limitations to teaching have changed in all the Nordic countries, and this change is strongly related to changes in achievements. In Norway, more in depth studies of this have been published, where findings point to a negative development over time; students got less sleep and were more often tired in school in 2019 than in 2015 (Kaarstein & Nilsen, 2021; Vik et al., 2022a, 2022b). Moreover, this explained changes in achievement in mathematics and science (ibid). Further, the decrease in how often students spoke Norwegian at home explained the decreased achievements of eighth grade science students from 2015 to 2019 (Lehre & Nilsen, 2021). Students' conceptual understanding in science decreased during this time period (Lehre et al., 2021). The decreased science achievement in eighth grade was further related to more students being uninterested and interrupting the instruction, and more students lacking prior knowledge (Kaarstein & Nilsen, 2021), as well as a less safe school environment (Nilsen et al., 2022). The changes in the students' habits and composition were confirmed by previous research that found: a negative development of school environment (Wendelborg et al., 2020); increased challenges related to sleep and nutrition (Vik et al., 2022a, 2022b); and an increase in the number of minority students (Sentralbyrå, 2023). The findings from Norway from 2015 to 2019, support the findings in the present study from 2011 to 2019, and may point to effects of negative developments of limitations to teaching being associated with changes in achievements over time rather than over grade. Moreover, two of the three timepoints (i.e., 2015 and 2019) included students from the same grade.

However, further in-depth studies, preferably with longitudinal designs, are needed to confirm these findings, especially in Denmark, Finland, and Sweden.

7.6.2   Limitations, Reliability, and Validity

The method used in the present study is more robust than analyses of one cycle of TIMSS, and more robust than comparing results from separate analyses of each cycle. Still, this trend study is based on cross-sectional data, and hence no causal inferences can be made. The validity is low for some constructs, due to changes of items in the construct over time. This is especially the case for cognitive activation in mathematics.

Since Norway changed their target grade in 2015, it is hard to interpret whether changes in teacher practices are related to changes in achievement over time or over grades, or both. This weakens the inferences related to Norway. However, as pointed out earlier, previous analyses for Norway from 2015 to 2019 regarding limitations to teaching, support our findings on this in the present study.

7.6.3   Concluding Remarks, Contributions, and Implications

The main aim of the present study was to examine the relationship between changes in teacher practices and changes in achievement over time. The findings indicate that teaching quality matters for students, that clarity of instruction decreased somewhat in all countries except in Finland where it increased, and that these changes are related to changes in achievement. However, changes in teaching quality only explain a minor part of the changes in achievement. The increased limitations on teaching quality do, however, explain a large part of the changes in achievement in all countries and in both subject domains. This finding is a serious one and points to a need for further research into why student composition and students' habits in terms of sleep and hunger at school have changed in the Nordic countries and what consequences this may have.

This study contributes to the field of research on teacher practices and teaching quality by confirming previous research from other countries (e.g. Baumert et al., 2010; Charalambous & Praetorius, 2020; Darling-Hammond, 2000) in the context of the Nordic countries. It further contributes to educational policy, stakeholders and practice, and points to the need for helping students in terms of their interest in mathematics and science, avoiding hungry students at school, and helping students who lack the necessary language skills. Parents may help their children get sufficient sleep, and researchers may want to further examine the impact of avoiding social media and gaming on sleep deprivation (Vik et al., 2022b).

Appendices

Appendix 1 How Changes in Teacher Practices Are Related to Changes in Science Achievement Over Time

	The effect of the predictor on science achievement				The effect of time on predictor				The effect of time on science achievement (direct effect)				Indirect effect			
	SWE	NOR	FIN	DEN	SWE	NOR	FIN	DEN	SWE	NOR	FIN	DEN	SWE	NOR	FIN	DEN
Clarity of instruction	– 1.7	6.9**	14.8**	2.6	– 0.1**	– 0.1**	0.0	0.1**	5.2**	22.8**	– 6.5**	3.0**	0.1	– 0.7**	0.3	– 0.1
Cognitive activation	50.3**	21.0**	7.0*	16.2	– 0.0	0.1**	0.0	0.0	6.1**	20.3**	– 6.4**	– 2.7	– 0.8*	1.8**	0.1	– 0.4
Limit	13.6**	10.0**	20.4**	13.1**	– 0.5**	– 0.3**	– 0.4**	– 0.5**	11.8**	25.2**	2.7	4.0*	– 6.5**	– 3.1*	– 9.0**	– 7.1**
Use of homework	– 14.2	3.3	6.2	3.1	0.0	– 0.1	– 0.1**	0.0	5.8**	22.3**	– 6.05	– 3.1*	– 0.5	– 0.2	– 0.2	0.1
Homework minutes	– 0.6*	– 0.2	0.19*	NC	– 1.1*	– 0.1	– 3.8**	NC	3.1	22.6**	5.16**	NC	0.6	0.0	– 0.7*	NC

Note NC means that the model did not converge, * denotes $p < 0.1$, ** denotes $p < 0.05$

Appendix 2 How Changes in Teacher Practices Are Related to Changes in Mathematics Achievement Over Time

	The effect of the predictor on math achievement				The effect of time on predictor				The effect of time on math achievement (direct effect)				Indirect effect			
	SWE	NOR	FIN	DEN	SWE	NOR	FIN	DEN	SWE	NOR	FIN	DEN	SWE	NOR	FIN	DEN
Clarity of instruction	16.6**	28.6**	28.4**	37.2**	0.1**	0.1**	0.1**	0.1**	12.4**	25.7**	6.8**	− 5.3**	1.1**	3.1**	0.8**	0.9*
Cognitive activation	− 2.4	1.4	− 6.9	− 9.3	0.1*	0.3**	0.0	0.0	11.5**	23.0**	6.1**	− 6.1**	− 0.2	− 0.5	0.0	− 0.1
Limit	12.8**	10.4*	20.3**	12.1**	− 0.5**	− 0.3**	− 0.4**	− 0.5**	17.5**	25.7**	0.0	0.3	6.2**	3.2*	9.0**	6.4**
Use of homework	− 12.5	6.1	7.2	− 2.3	0.0	− 0.1	− 0.1**	0.0	11.8**	22.9**	5.8**	− 6.1**	− 0.5	− 0.3	− 0.3	− 0.1
Homework minutes	− 0.3**	0.0	− 0.1	0.2*	− 4.1**	0.4	− 7.0**	− 4.5**	11.5**	19.0**	− 7.9**	− 12.4**	1.4*	0.0	0.9	− 0.9

References

Baumert, J., Kunter, M., Blum, W., Brunner, M., Voss, T., Jordan, A., Klusmann, U., Krauss, S., Neubrand, M., & Tsai, Y.-M. (2010). Teachers' mathematical knowledge, cognitive activation in the classroom, and student progress. *American Educational Research Journal, 47*(1), 133–180. https://doi.org/10.3102/0002831209345157

Charalambous, C. Y., & Praetorius, A.-K. (2020). Creating a forum for researching teaching and its quality more synergistically. *Studies in Educational Evaluation, 67*, 100894. https://doi.org/10.1016/j.stueduc.2020.100894

Darling-Hammond, L. (2000). Teacher quality and student achievement. *Education Policy Analysis Archives, 8*. https://doi.org/10.14507/epaa.v8n1.2000

Fan, H., Xu, J., Cai, Z., He, J., & Fan, X. (2017). Homework and students' achievement in math and science: A 30-year meta-analysis, 1986–2015. *Educational Research Review, 20*, 35–54. https://doi.org/10.1016/j.edurev.2016.11.003

Fauth, B., Decristan, J., Rieser, S., Klieme, E., & Büttner, G. (2014). Student ratings of teaching quality in primary school: Dimensions and prediction of student outcomes. *Learning and Instruction, 29*, 1–9. https://doi.org/10.1016/j.learninstruc.2013.07.001

Fernández-Alonso, R., & Muñiz, J. (2021). Homework: Facts and fiction. In T. Nilsen, A. Stancel-Piątak, & J.E. Gustafsson (Eds.), *International handbook of comparative large-scale studies in education*. Springer International Handbooks of Education. https://doi.org/10.1007/978-3-030-38298-8_40-1

Gustafsson, J.-E. (2013). Causal inference in educational effectiveness research: A comparison of three methods to investigate effects of homework on student achievement 1. *School Effectiveness and School Improvement, 24*(3), 275–295. https://doi.org/10.1080/09243453.2013.806334

Gustafsson, J.-E., & Nilsen, T. (2022). Methods of causal analysis with ILSA data. In T. Nilsen, A. Stancel-Piątak, & J.-E. Gustafsson (Eds.), *International handbook of comparative large-scale studies in education: Perspectives, methods and findings* (pp. 1–28). Springer. https://doi.org/10.1007/978-3-030-38298-8_56-1

Kaarstein, H., & Nilsen, T. (2021). Lærerkompetanse, undervisningskvalitet og naturfagprestasjoner fra TIMSS 2015 til TIMSS 2019 [Teacher competence, teaching quality and science performance from TIMSS 2015 to TIMSS 2019]. In *Med blikket mot naturfag: Nye analyser av TIMSS 2019-data og trender 2015–2019 [Focused on science: new analyses of TIMSS 2019 data and trends from 2015–2019]* (pp. 183–206). Universitetsforlaget. https://doi.org/10.18261/978821 5045108-2021-08

Klieme, E., Pauli, C., & Reusser, K. (2009). The pythagoras study: Investigating effects of teaching and learning in Swiss and German mathematics classrooms. *The power of video studies in investigating teaching and learning in the classroom*, 137–160.

Lehre, A.-C. W., Frønes, T. S., & Kaarstein, H. (2021). TIMSS 2019: Hverdagsspråk og naturfaglig diskurs i elevenes svar på åpne oppgaver [TIMSS 2019: Everyday language and scientific discourse in students' responses to open-ended questions]. In *Med blikket mot naturfag: Nye analyser av TIMSS 2019-data og trender 2015–2019 [Focused on science: new analyses of TIMSS 2019 data and trends from 2015–2019]* (pp. 73–102). Universitetsforlaget. https://doi.org/10.18261/9788215045108-2021-04

Lehre, A.-C. W., & Nilsen, T. (2021). Språk i hjemmet og naturfagprestasjoner fra TIMSS 2015 til TIMSS 2019 [Language at Home and Science Performance from TIMSS 2015 to TIMSS 2019]. In *Med blikket mot naturfag: Nye analyser av TIMSS 2019-data og trender 2015–2019 [Focused on science: new analyses of TIMSS 2019 data and trends from 2015–2019]* (pp. 165–182): Universitetsforlaget. https://doi.org/10.18261/9788215045108-2021-07

Marder, J., Thiel, F., & Göllner, R. (2023). Classroom management and students' mathematics achievement: The role of students' disruptive behavior and teacher classroom management. *Learning and Instruction, 86,* 101746. https://doi.org/10.1016/j.learninstruc.2023.101746

Mullis, I. V. S., Martin, M. O., Foy, P., Kelly, D., & Fishbein, B. (2020). *TIMSS 2019 International results in mathematics and science.* TIMSS & PIRLS International Study Center, Boston College. https://timssandpirls.bc.edu/timss2019/international-results/

Murnane, R. J., & Willett, J. B. (2010). *Methods matter: Improving causal inference in educational and social science research.* Oxford University Press.

Muthén, L. K., & Muthén, B. O. (1998–2017). *Mplus user's guide* (8th ed.). Muthén & Muthén.

Nilsen, T., & Gustafsson, J.-E. (Eds.). (2016). *Teacher quality, instructional quality and student outcome. Relationships across countries, cohorts and time* (Vol. 2). Springer https://doi.org/10.1007/978-3-319-41252-8

Nilsen, T., Kaarstein, H., & Lehre, A.-C. (2022). Trend analyses of TIMSS 2015 and 2019: School factors related to declining performance in mathematics. *Large-Scale Assessments in Education, 10*(1), 15. https://doi.org/10.1186/s40536-022-00134-8

Nilsen, T., Scherer, R., & Blömeke, S. (2018). The relation of science teachers' quality and instruction to student motivation and achievement in the 4th and 8th grade: A Nordic. *Northern Lights on TIMSS and PISA 2018, 61.* https://www.norden.org/en/publication/northern-lights-timss-and-pisa-2018

Olsen, R. V., & Björnsson, J. K. (2018). Fødselsmåned og skoleprestasjoner [Month of birth and school performance]. In J. K. Björnsson & R. V. Olsen (Eds.), *Tjue år med TIMSS og PISA i Norge. Trender og nye analyser* [Twenty years with TIMSS and PISA in Norway. Trends and new analyses.]. Universitetsforlaget. https://doi.org/10.18261/9788215030067-2018

Praetorius, A.-K., Klieme, E., Herbert, B., & Pinger, P. (2018). Generic dimensions of teaching quality: The German framework of Three Basic Dimensions. *ZDM, 50,* 407–426. https://doi.org/10.1007/s11858-018-0918-4

Senden, B., Nilsen, T., & Blömeke, S. (2022). Instructional quality: A review of conceptualizations, measurement approaches, and research findings. In M. Blikstad-Balas, K. Klette, & M. Tengberg (Eds.), *Ways of analyzing teaching quality: Potentials and pitfalls* (pp. 140–172). Scandinavian University Press. https://doi.org/10.18261/9788215045054-2021-05

Senden, B., Nilsen, T., & Teig, N. (2023). The validity of student ratings of teaching quality: Factorial structure, comparability, and the relation to achievement. *Studies in Educational Evaluation, 78,* 101274. https://doi.org/10.1016/j.stueduc.2023.101274

Sentralbyrå, S. (2023). Befolkningsendringer, etter innvandringskategori 2011–2021. *Statistics Norway.* https://www.ssb.no/statbank/table/10516/

Teig, N., Bergem, O. K., Nilsen, T., & Senden, B. (2021). Gir utforskende arbeidsmåter i naturfag bedre læringsutbytte? [Does inquiry-based teaching practice in science provide better learning outcomes?]. In T. Nilsen & H. Kaarstein (Eds.), *Med blikket mot naturfag [A view towards science]* (pp. 46–72). Universitetsforlaget. https://doi.org/10.18261/9788215045108-2021-03

Teig, N., Scherer, R., & Nilsen, T. (2018). More isn't always better: The curvilinear relationship between inquiry-based teaching and student achievement in science. *Learning and Instruction, 56,* 20–29. https://doi.org/10.1016/j.learninstruc.2018.02.006

Trautwein, U. (2007). The homework–achievement relation reconsidered: Differentiating homework time, homework frequency, and homework effort. *Learning and Instruction, 17*(3), 372–388. https://doi.org/10.1016/j.learninstruc.2007.02.009

Vik, F. N., Nilsen, T., & Øverby, N. C. (2022a). Aspects of nutritional deficits and cognitive outcomes—Triangulation across time and subject domains among students and teachers in TIMSS. *International Journal of Educational Development, 89*, 102553. https://doi.org/10.1016/j.ijedudev.2022.102553

Vik, F. N., Nilsen, T., & Øverby, N. C. (2022b). Associations between sleep deficit and academic achievement—triangulation across time and subject domains among students and teachers in TIMSS in Norway. *BMC Public Health, 22*(1), 1790. https://doi.org/10.1186/s12889-022-14161-1

Wendelborg, C., Dahl, T., Røe, M., & Buland, T. H. (2020). *Elevundersøkelsen 2019: Analyse av Utdanningsdirektoratets brukerundersøkelser* [*Student Survey 2019: Analysis of the Directorate of Education's user surveys*]. NTNU Samfunnsforskning. https://www.udir.no/tall-og-forskning/finn-forskning/rapporter/elevundersokelsen-2019-hovedrapporten/

Trude Nilsen is a research professor at the University of Oslo. She is a leader of Strand 1 at CREATE - Centre for Research on Equality in Education, a leader of the research group LEA, and of funded research projects. She has been engaged as an international external expert for IEA's TIMSS and for OECD's TALIS studies. Her research focuses on teaching quality, educational equality, school climate, and applied methodology including causal inferences.

Jan-Eric Gustafsson (born 1949) is a professor emeritus of education at the University of Gothenburg, Sweden. His research has primarily focused on basic and applied topics within the field of educational psychology, such as individual prerequisites for education, effects of resources, and educational outcomes at individual and system levels. He was elected member of the Royal Swedish Academy of Sciences in 1993, elected Corresponding Fellow of the British Academy in 2021, and received an honorary doctorate from the University of Oslo in 2017.

Chapter 8
Equality in Content Coverage in the Nordic Countries?

Sigrid Blömeke ⓘ

8.1 Introduction

Systematic inequalities in student achievement are of concern in all Nordic countries. The state of research reveals, for example, substantial gaps in mathematics achievement between children from families with high and low socioeconomic status (SES) (Sandsør et al., 2021). It is rarely questioned that such inequalities are caused by different family backgrounds of students. However, these provide students with different prerequisites for succeeding at school and different support during schooling. Which role schooling itself plays in contributing to socioeconomic inequalities in student achievement is largely an open question, although such research could point to malleable factors that either contribute to or could have the potential to mitigate inequalities.

A feature of schooling that is discussed very little in this context is the content taught to students, although students' chance to learn a subject depends on the content they are exposed to (Schmidt & McKnight, 2012). The role of "content coverage" defined as the extent to which "students have had the opportunity to study a particular topic" (Husén, 1967, p. 162) for socioeconomic inequalities in student achievement in the Nordic countries has received limited attention in the existing research. In principle, content coverage should largely be the same for all students during compulsory schooling. Equal opportunities in a "School for All" has been a cornerstone of the educational philosophy in the Nordic countries since after the Second World War (Blossing et al., 2014). Students of all backgrounds should learn

Sigrid Blömeke: Deceased

S. Blömeke (✉)
Centre for Educational Measurement (CEMO) and Centre for Research on Equality in Education (CREATE), University of Oslo, Forskningsparken, Gaustadalleen 21, Oslo N-0373, Norway

© The Author(s) 2024
N. Teig et al. (eds.), *Effective and Equitable Teacher Practice in Mathematics and Science Education*, IEA Research for Education 14,
https://doi.org/10.1007/978-3-031-49580-9_8

together, and education should be free and public (Bostad & Solberg, 2022). During the 1950s to 1960s, all Nordic countries introduced comprehensive school systems without formal tracking and since then, private schools have only been allowed to a very limited extent (Antikainen, 2006). Furthermore, national curricula should further ensure that all students receive equal opportunities to learn independently from where they go to school or whom they are.

However, whether the reality of equal opportunities is in line with this ideal is an open question since there is little research in this area. One may even argue that there is no longer a "Nordic" model in education (Lundahl, 2016). The education systems both in Denmark, Finland, Norway, and Sweden—the Nordic countries included in the present study—have developed (to varying degrees) towards decentralization and school choice. Among these countries, Sweden represents an extreme case in these developments. In line with this development, studies have provided evidence for increasing inequalities between schools with respect to educational resources and teacher quality and in addition increasing inequalities in student achievement (Buchholtz et al., 2020). This development may also include a risk of inequalities in opportunities to learn in terms of content coverage—with the worst-case consequence that schooling may contribute to inequalities instead of mitigating them.

On the other hand, the mission of public education in all Nordic countries is still—more or less explicitly—to increase social mobility and to contribute to equal societal chances of every student (Opheim, 2004). Schools are therefore required to compensate for student disadvantages caused by differences in their family background, with the minimization of social differences as the objective (Imsen & Volckmar, 2014). To use Denmark as an example, where the law states: "Folkeskolen skal mindske betydningen af social baggrund i forhold til faglige resultater" [The public school should reduce the significance of social background in relation to academic outcomes]. (cf. Reimer et al., 2018). Similar regulations exist in all Nordic countries. Moreover, preparing teachers for supporting students in accordance with their individual needs has become the focus of educational policy, as heterogeneity has increased in the Nordic as in all other European countries. With respect to content coverage, such a development could mean exactly the opposite of the concerns mentioned above, namely providing additional content coverage for students with disadvantaged backgrounds.

Based on the 2019 data from the Trends in Mathematics and Science Studies (TIMSS), the present study examines whether inequalities in the content taught to students exist in Denmark, Finland, Norway, or Sweden and whether these inequalities are related to student SES. To address this aim, this chapter first describes any variation in student SES and the mathematics content taught to students in the four countries. Secondly, it examines whether the mathematics content taught varies by students' SES. Finally, this study formally tests the joint effect of SES and content coverage on students' achievement in mathematics.

8.2 Conceptual Framework

8.2.1 Socioeconomic Inequalities

The socioeconomic status of students can be measured in different ways, depending on the theoretical understanding of this construct and the data available. Based on Bourdieu (1986), sociological theories often distinguish between cultural, economic, and social capital. In principle, it is possible to assess all of these dimensions with TIMSS data, and TIMSS also provides an index including all dimensions. However, when it comes to relations to educational outcomes, research has provided evidence that cultural capital might be more important than economic or social capital. These results apply to the Nordic countries with their still relatively flat societal structure in an international context and free public education (Møllegaard & Meier Jæger, 2015). Moreover, the TIMSS indicators of economic capital have large proportions of missing data in several countries, which hampers the representativeness of the results. The validity of the social-capital indicators and the validity and reliability of the composite index have also been questioned (Ye et al., 2021). In this chapter, we therefore focus on students' cultural capital by utilizing the variable "number of books at home".

Wiberg and Rolfsman (2021) provided evidence for grade eight students in Sweden that this TIMSS indicator is a valid representation of students' backgrounds. These researchers had the unique opportunity to link students' TIMSS data to official Swedish register data via students' personal identification numbers, and they found that both provided the same information. However, it has been argued that younger students in grade four might not be fully capable of reliably providing this information (Brese & Mirazchiyski, 2013). TIMSS provides information about the number of books at home, both from students and from the parents' responses. Newer evidence with non-TIMSS samples supports greater validity of this source (Heppt et al., 2022). However, the response rate of parents in TIMSS has turned out to be rather low. Although there is no reason to believe that the respective limitations of either the student or the parent indicator would play out differently across the Nordic countries (which means that they do not bias the comparison; Reimer et al., 2018), we address the concerns regarding validity and representativeness by combining parents' and students' reports about the number of books at home. By combining these sources, we are able to mutually compensate for the respective weaknesses.

8.2.2 Content Coverage

TIMSS distinguishes between the intended curriculum prescribed at the system level, the implemented curriculum at the classroom level, and the attained curriculum as students' learning outcomes (Mullis & Martin, 2017). In line with these distinctions, the intended curriculum can be assessed by analyzing national curricula, the implemented curriculum by asking teachers what they have taught, and the attained curriculum by testing students on their achievement. Most of the curriculum research has made use of the first approach, since there has been a general lack of large-scale information on the specific topics taught by teachers (Schmidt et al., 2015). The international large-scale assessments close this gap. In the present study, we use teachers' reports about the topics they covered in mathematics either taught this year or the year before, versus not yet covered to assess the implemented curriculum in Denmark, Finland, Norway, and Sweden. Mathematics has traditionally been a domain accessible for international comparisons since a joint understanding about content areas and topics exists. The list of topics has been developed by content experts, and TIMSS has conducted a range of validity studies to make sure that teachers' reports provide meaningful information.

8.2.3 Inequalities in Content Coverage

There are different ways to examine inequalities in content coverage: One can, firstly, look at the variance in topics taught to students with the underlying hypothesis that the larger the variance, the less equal opportunities exist. Such a study can be done at the individual student level and the aggregated (classroom or school) level. The latter would indicate so- called composition effects, with the underlying hypothesis that there are additional SES effects on content coverage if more students of the same type of background are present in a class. Evidence in this respect exists for societies with highly fragmented neighborhoods or school systems with tracking. In the Nordic countries, however, with their comprehensive school systems up to the end of lower-secondary school, and less fragmented societies, it is an open question whether composition effects exist.

Secondly, one can examine the differences in content coverage by students' socioeconomic background. Given that the state of research points to particular advantages or disadvantages at the ends of the distributions, the student groups should be defined rather than fine-grained. We distinguish between five groups, each representing one quintile of a country's SES distribution (20%). This approach takes into consideration absolute SES differences between the four Nordic countries and potential problems with measurement invariance of the SES measure. Low- or high-SES background is defined relative to a country's distribution, instead of using one fixed SES threshold across countries. The results can thus be compared meaningfully, relative to countries' SES distribution (Ye et al., 2021).

8.2.4 Socioeconomic Inequalities in Content Coverage and Student Achievement

Socioeconomic inequalities in student achievement due to different family backgrounds are a well-known phenomenon. If, in addition, inequalities in content coverage exist by socioeconomic background and if content coverage is related to student achievement, the inequalities could exacerbate the inequalities stemming from student background. However, the relationship between SES, content coverage, and student achievement is largely an open question with respect to the Nordic countries. This means that a risk exists of exaggerating SES effects and masking the role of schooling if the models do not include potential inequalities in the topics taught to students and just include their background. In line with Schmidt et al. (2015), we test a mediation model, where student SES has an effect on achievement but also on the content they are exposed to.

This is not the first study to apply the idea of effects of family background on student achievement mediated by inequalities in education. However, while this is a common approach in the USA, research on this topic in the Nordic countries is scarce. It is not far-fetched to assume that this is related to the general belief of equal opportunities for all children. We found only one study (Rolfe et al., 2021) where this approach was applied to grade eight students in a large set of countries including Finland, Norway, and Sweden. The present study with grade four students will therefore enter largely uncharted territory.

8.2.5 Research Questions

The following research questions guide this study:

1. How much variance in student SES exists in the Nordic countries?
2. How much variance in content coverage in mathematics exists in the Nordic countries? Are there differences across subdomains?
3. Are groups of students with different SES backgrounds exposed to different amounts of topics taught in the Nordic countries? Are there differences across content domains?
4. How much variance in mathematics achievement exists in the Nordic countries? Do groups of students with different SES backgrounds have different levels of mathematics achievement?
5. Are the effects of student SES on achievement mediated by inequalities in content coverage?

Table 8.1 Country-specific data

Country	No. students	No. classes/teachers	Average class size
Denmark	3227	195	16.5
Finland	4730	321	14.7
Norway	3951	230	17.2
Sweden	3965	224	17.7

8.2.6 Methodology

Sample and Procedure

We utilized TIMSS 2019 data, in which one or more intact classes were selected from randomly sampled schools via a two-stage stratified cluster sampling design (Mullis & Martin, 2017). The raw data included, in most cases, one teacher per class. For brevity, we also selected randomly one teacher per class in cases where there were several teachers per class. The sample comprises fourth-grade students ($N =$ 15,873) and their teachers ($N = 970$) from Denmark, Finland, Norway, and Sweden (see Table 8.1 for the exact size of the samples). For Denmark and Norway, the samples did not fully meet the required response rates, with teacher data available for at least 70 percent but less than 85 percent of the students. Therefore, the results have to be used with great care. Moreover, Norway collected data from grade five, while all other Nordic countries collected data from grade four.[1]

Measures

Socioeconomic status of students

Student SES was assessed with the item "about how many books are there in your home?" with five categories: 0–10, 11–25, 26–100, 101–200, and more than 200 books. Students and parents provided information to the item. Reliability was moderate to good on the within level, and very good on the between level, in all four Nordic countries: On the within-level, the two measures correlated at around 0.50–0.60 and on the between level at around 0.90. The individual student SES variable (i.e., differences within classrooms) was created by averaging the information from students and parents. In case one source was missing, the other one was used. The class composition variable (i.e., differences between classrooms) was created by aggregating the individual information on the between level.

---

[1] Since TIMSS 2015, Norway has shifted the target population from grade four to grade five. In Norway, formal education starts at age six with grade one, which means that students in grade four are usually younger compared to their counterparts in many other participating countries. By shifting the target grade, Norway aimed to match more closely the age of students being tested in other countries.

Content coverage

We used, firstly, the percentage of students taught the corresponding topics. This is an index provided by TIMSS where teachers' responses "mostly taught before this year" or "mostly taught this year" were coded as 1 and then averaged across all items belonging to the respective area, such as number in mathematics. Prompted by "the following list includes the main topics addressed by the TIMSS mathematics test. Choose the response that best describes when the students in this class have been taught each topic", teachers had to rate 17 mathematics items respectively that covered three content domains:

- Seven items for number, e.g., "concepts of whole numbers, including place value and ordering".
- Seven items for measurement and geometry, e.g., "solving problems involving length, including measuring and estimating".
- Three items for data, e.g., "reading and interpreting data from tables, pictographs, bar graphs, line graphs, and pie charts".

Student Achievement

The TIMSS 2019 test for students in fourth grade included 175 mathematics items resulting in 190 score points requiring students to use knowing (33%), application (43%), and reasoning (24%) skills (Mullis & Martin, 2017). Forty-six percent of the score points came from selected-response and 54% from constructed-response items. The test covered three content domains, with 47% devoted to number, 31% devoted to measurement and geometry, and 23% devoted to data. Since our objective is to examine the effects of content coverage in the respect subareas, we will also use the mathematics subscores provided by TIMSS for number, measurement and geometry, and data.

8.2.7 Statistical Analyses

TIMSS 2019 data are available from the International Association for the Evaluation of Education Achievement (IEA) database[2]. Statistical modelling was conducted with Mplus version 8.8 (Muthén & Muthén, 2017). We used all five plausible values provided by TIMSS based on students' responses to the test items and conditioned all available background data (Laukaityte & Wiberg, 2017). Maximum likelihood estimation with robust standard errors (MLR) was applied, and missing data were handled with the full information maximum likelihood (FIML). The Nordic countries were included simultaneously by implementing a multigroup model. To evaluate the

---

[2] TIMSS 2019 data can be accessed from the IEA database (https://timss2019.org/international-dat abase/).

model fit, we used common guidelines with respect to a series of fit indices (Marsh et al., 2005). Weights were used in line with Rutkowski et al. (2010).

8.3 Results

RQ 1: Variance in student SES in the Nordic countries

In all four Nordic countries, there were on average enough books at home to fill one bookcase, which means roughly between 26 and 100 books (see Table 8.2). There were significantly fewer books at home in Denmark than in the other three countries. Student background was more heterogenous in Sweden than in Denmark and Norway, and in all three countries more heterogenous than in Finland.

RQ 2: Variance in mathematics content coverage in the Nordic countries

Mathematics teachers in Sweden reported significantly fewer topics taught to grade four students than teachers in Norway, and these again were significantly fewer than teachers in Denmark and Finland (see Table 8.3).

Table 8.2 Mean student SES based on parent and student information (standard error)

Country	Min–max	M (SE)	Variance
Denmark	0–4	2.00 (0.02)	1.24
Finland	0–4	2.20 (0.01)	1.00
Norway	0–4	2.22 (0.02)	1.25
Sweden	0–4	2.20 (0.02)	1.34

Notes 0 = None or very few (0–10 books), 1 = Enough to fill one shelf (11–25 books), 2 = Enough to fill one bookcase (26–100 books), 3 = Enough to fill two bookcases (101–200 books), 4 = Enough to fill three or more bookcases (more than 200)

Table 8.3 Mean percentages of students taught mathematics topics (standard error)

Country	Mathematics	Number	Geometry	Data
Denmark	77 (1.1)	84 (1.3)	75 (1.5)	62 (2.8)
Finland	77 (1.0)	93 (0.8)	69 (1.5)	58 (2.5)
Norway	70 (1.5)	77 (1.7)	60 (2.2)	78 (3.7)
Sweden	65 (1.5)	72 (1.4)	57 (2.4)	68 (3.7)

Note Source: Mullis et al. (2020)

In Denmark and Finland, number topics are very pronounced in relation to the average amount of mathematics topics, while students are less exposed to data topics. The pattern is different in Sweden and Norway where data is very pronounced in relation to the average amount of mathematics topics, while students are less exposed to geometry topics.

The between-classroom variance reveals that there is large variation in the amount of number topics taught in Denmark, Norway and Sweden, while the amount of number topics taught in Finland are relatively homogenous across different classrooms (see Table 8.3 and Fig. 8.1). The variance is in all countries larger when it comes to geometry and measurement, and again even larger when it comes to data.

RQ 3: Inequalities in mathematics content coverage in the Nordic countries

With respect to the content covered in Denmark, there is the same tendency visible in all three content domains (see Fig. 8.2): The number of topics taught to grade four students increases by students' SES.

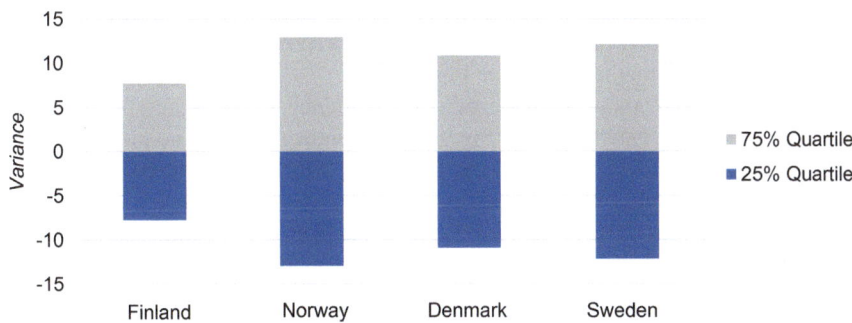

Fig. 8.1 Between-classroom content coverage in the Nordic countries

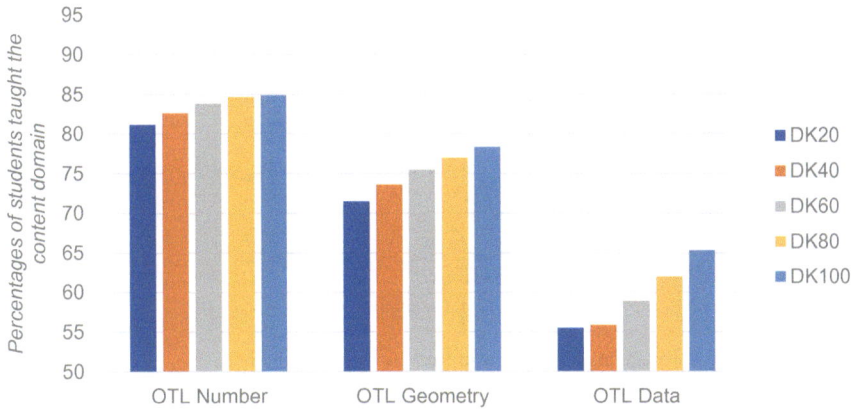

Fig. 8.2 Content coverage in mathematics in Denmark by SES quintiles

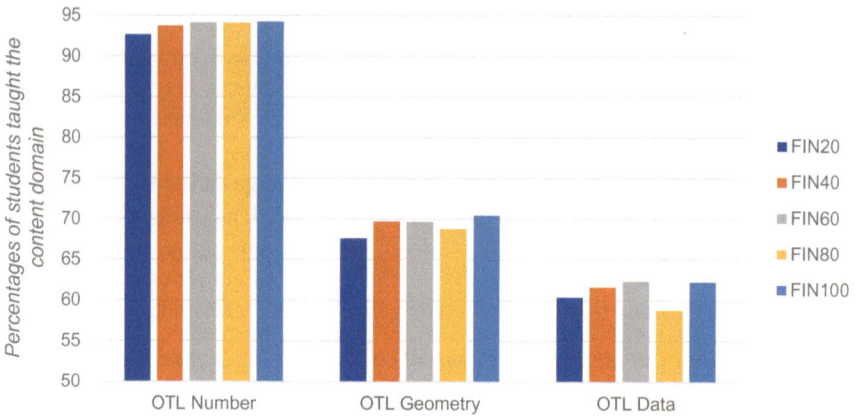

Fig. 8.3 Content coverage in mathematics in Finland by SES quintiles

The data for Finland reveal a relatively balanced SES profile with a slight tendency of covering more topics at the upper end of the SES specter. This applies to all three content domains of mathematics (see Fig. 8.3).

With respect to Norway, there seems to be a tendency that more number topics are taught to higher SES students (see Fig. 8.4). In contrast, there is no clear tendency for geometry or data—if at all, there is a tendency of more content coverage at the lower and upper ends of the SES specter.

Finally, there is no clear pattern for Sweden (see Fig. 8.5). If at all, one could infer a tendency of fewer topics taught to high-SES students in all three subdomains.

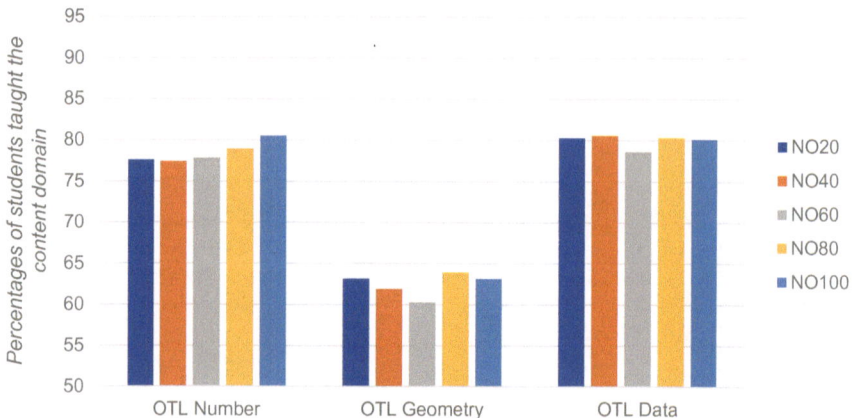

Fig. 8.4 Content coverage in mathematics in Norway by SES quintiles

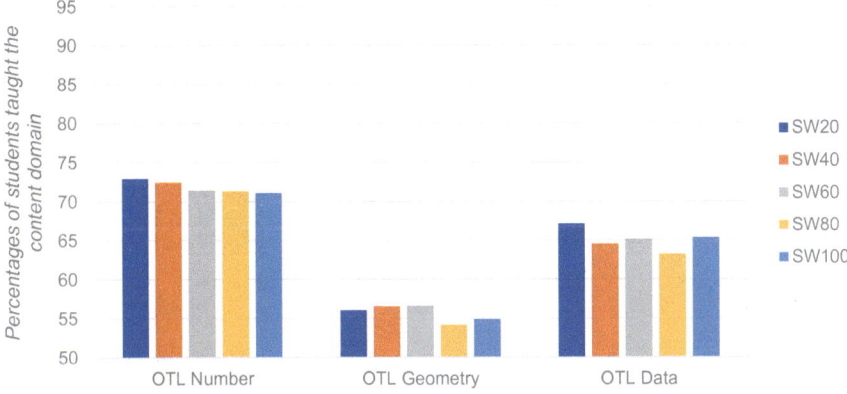

Fig. 8.5   Content coverage in mathematics in Sweden by SES quintiles

RQ 4: Mathematics achievement in the Nordic countries

The average achievement level in mathematics was highest in Norway (grade five students) and lowest in Denmark and Sweden, with Finland in the middle (see Table 8.4). All four Nordic countries typically score lower in the number content domain while they do better in geometry, relative to their average mathematics scores. Denmark stands out with the largest differences between geometry and average mathematics achievement, whereas no such difference was found in Sweden.

If one splits up the variance in mathematics achievement into within- and between-school variance, the data reveal that Finland, Sweden, and Norway have larger between-school variance than Denmark (see Fig. 8.6).

The distribution of mathematics achievement per student group reveals that SES is clearly related to achievement in all Nordic countries (see Fig. 8.7). It is particularly noteworthy how similarly large the gaps are between students from the lowest and the highest quintiles of the SES distributions and how evenly the gaps increase by percentile.

Table 8.4   Student achievement in mathematics and its content domains

Country	Mathematics	Number	Diff.	Geometry	Diff.	Data	Diff.
Denmark	525 (1.9)	518 (2.1)	− 7 (1.1)	536 (2.4)	12 (1.8)	525 (2.3)	1 (1.5)
Finland	532 (2.3)	528 (2.3)	− 4 (1.0)	538 (3.0)	6 (2.2)	534 (2.8)	2 (1.8)
Norway (5)	543 (2.2)	540 (2.0)	− 3 (1.0)	546 (2.8)	4 (1.5)	547 (3.2)	4 (2.4)
Sweden	521 (2.8)	517 (2.9)	− 4 (1.4)	521 (3.4)	0 (1.7)	527 (3.5)	6 (1.8)

Source Mullis et al. (2020)

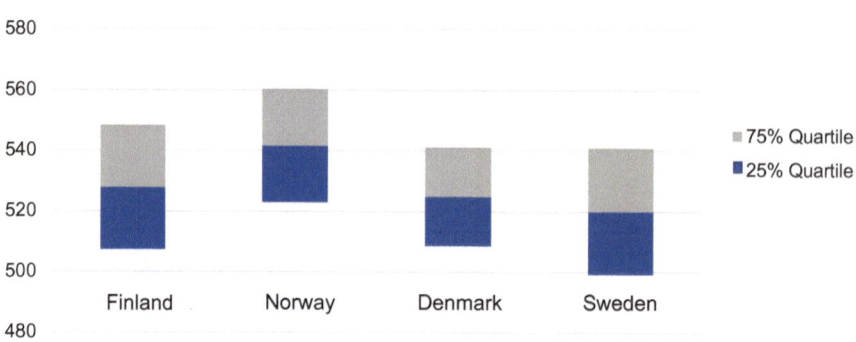

Fig. 8.6 Between-classroom variance in mathematics achievement in the Nordic countries

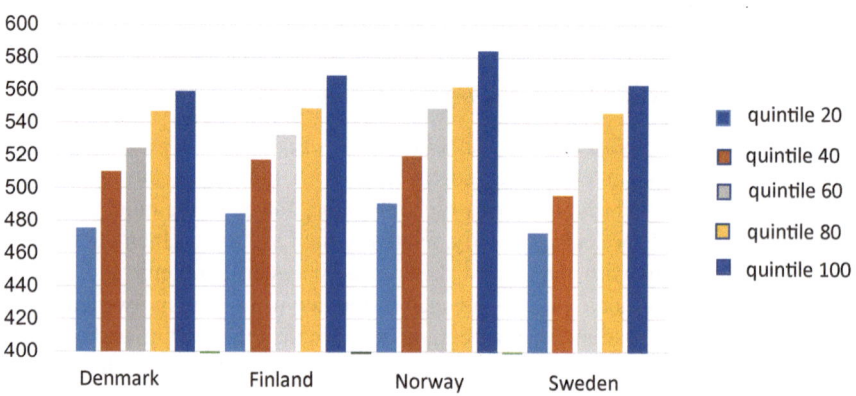

Fig. 8.7 Distribution of mathematics achievement by SES quintiles

RQ 5: Effects of SES and content coverage on mathematics achievement

Results from the mediation model that jointly estimates the effects of SES and content coverage on grade four students' achievement in mathematics are documented in Table 8.5.

On the within-classroom level, achievement has little variation that is significantly related to student SES in all three content domains for mathematics and in all four countries (as displayed in Fig. 8.7).

On the between-classroom level, students' achievement is strongly related to SES in all domains and countries as well. Such a composition effect indicates an additional effect of how similar students' SES is in a class *beyond* individual SES effects, meaning the higher a class's average SES is, the higher its average achievement is. Vice versa, the lower students' average SES is in a class, the lower their achievement is. Note the particularly strong composition effect of SES on achievement in all three domains in Finland.

Table 8.5 Multilevel multigroup mediation model of the relation between student SES, content coverage, and student achievement

Number

Fixed effects	DK			FIN			NO			SW		
	β	SE	p	β	SE	p	β	SE	p	β	SE	p
Within (N)	2810			4655			3143			3543		
Achieve on												
SES	0.28	0.02	*	0.30	0.02	*	0.29	0.03	*	0.31	0.03	*
Between (k)	195			320			230			224		
Achieve on												
OTL	0.29	0.01	*	0.17	0.10	*	0.39	0.15	*	0.13	0.10	Ns
SES_cl	0.37	0.10	*	0.60	0.09	*	0.27	0.15	*	0.40	0.13	*
OTL on												
SES_cl	0.17	0.10	*	0.10	0.07	ns	0.06	0.12	ns	− 0.14	0.09	Ns
SES_clind	0.05	0.03	ns	0.02	0.02	ns	0.02	0.04	ns	− 0.02	0.02	Ns

Geometry

Fixed effects	DK			FIN			NO			SW		
	β	SE	p	β	SE	p	β	SE	p	β	SE	p
Within (N)	2810			4655			3143			3543		
Achieve on												
SES	0.30	0.02	*	0.29	0.02	*	0.30	0.02	*	0.33	0.03	*
Between (k)	195			320			230			224		
Achieve on												
OTL	0.32	0.13	*	0.01	0.08	ns	0.15	0.13	ns	0.22	0.10	*
SES_cl	0.28	0.14	*	0.62	0.08	*	0.41	0.14	*	0.48	0.12	*
OTL on												
SES_cl	0.31	0.10	*	0.01	0.07	ns	0.10	0.13	ns	− 0.10	0.09	ns
SES_clind	0.10	0.05	*	0.00	0.00	ns	0.02	0.03	ns	− 0.02	0.02	ns

Data

Fixed effects	DK			FIN			NO			SW		
	β	SE	p	β	SE	p	β	SE	p	β	SE	p
Within (N)	2810			4655			3143			3543		
Achieve on												
SES	0.24	0.03	*	0.29	0.03	*	0.26	0.02	*	0.31	0.02	*
Between (k)	195			320			230			224		
Achieve on												
OTL	0.34	0.10	*	0.06	0.06	ns	0.18	0.08	ns	0.08	0.08	ns
SES_cl	0.30	0.11	*	0.59	0.08	*	0.39	0.08	*	0.34	0.15	*
OTL on												
SES_cl	0.15	0.10	ns	0.08	0.06	ns	− 0.16	0.09	ns	0.05	0.09	ns
SES_clind	0.05	0.04	ns	0.00	0.01	ns	− 0.03	0.03	ns	0.00	0.01	ns

Note * $p < 0.05$.

Student achievement is also significantly related to the content students have been exposed to. This applies to all three content domains in Denmark, while the degree varies in the other three countries from not significant at all to effects of medium size.

With respect to inequalities in content coverage, one needs to distinguish between countries and content domains of mathematics. In Finland, the data do not reveal systematic relations between student SES and the number of topics taught in the different domains. Consequently, no mediation effects of content coverage exist. This means that the slight upwards tendencies visible in Fig. 8.3 are not large enough to be regarded as systematic gaps in opportunities to learn.

As displayed in Fig. 8.2, Danish teachers have on average taught more topics on number and geometry to high-SES classrooms than to low-SES classrooms. The corresponding indirect effect on student achievement is significant in geometry. A similar tendency is visible in data but not large enough to be regarded as systematic.

The data for Norway and Sweden again reveal different results. Whereas most of the relations displayed in Figs. 8.4 and 8.5 are not strong enough to be statistically significant, Norwegian students from lower socioeconomic backgrounds are more often exposed to topics on data than students from higher SES backgrounds. A similar tendency exists in Sweden for number topics. However, none of the effects are large enough to state an additional mediation effect on student achievement.

8.4   Discussion

The Nordic countries strive to provide equal learning opportunities for all students. However, the reality looks quite different both in Denmark, Finland, Norway, and Sweden. All four countries have systematic gaps in grade four students' individual achievement in mathematics, and the gaps are with little variation and strongly relate to students' SES. The *causes* of these relations are difficult to identify. The present chapter may provide some preliminary answers based on cross-sectional data:

1. The variation in students' SES points to much larger socioeconomic differences in Sweden than in Denmark, Finland, and Norway. This result confirms earlier studies that the Swedish schools in Sweden have become more fragmented (Yang Hansen & Gustafsson, 2016). A number of reasons for this development exist. However, it is noteworthy that the relationship between achievement and SES is not stronger in Sweden than in the other three countries. One could interpret this as an indicator for schools' success in striving for providing equal opportunities (Trumberg et al., 2022).

2. Nevertheless, strong composition effects of SES on achievement exist in all Nordic countries. This means that individual SES effects are exacerbated by the similarity of students' SES background in a classroom: The more homogenous students are, the higher or lower a classes' achievement is, outside of individual differences. This applies to all three TIMSS content domains for mathematics in

Finland, a surprising result given most of the discussions in the literature where "success through equity" (Ahonen, 2021) is assumed. However, as pointed out by Yang Hansen and Gustafsson (2014) substantial differences in achievement between classrooms exist in Finland based on TIMSS 2011 data. Furthermore, Finnish researchers showed in their TIMSS 2019 national report (Vainikainen & Harju-Luukkainen, 2020), that a big proportion of this variance can be explained by classes where most of a school's students with special needs are compiled together (Sundqvist & Hannås, 2021) and through selective classes with an emphasis on language, arts, or sports where intake is based on an aptitude test (Kosunen & Seppänen, 2015). The strong composition effects of SES on achievement therefore indicate that equal opportunities to learn are a myth in the case of Finland but that tracking within schools related to student SES[3] exists.

3. With respect to content coverage, the data about the mathematics topics taught to grade four students reveal remarkably different amounts and profiles in the Nordic countries. A sharp distinction exists between Denmark and Finland on the one hand and Norway and Sweden on the other hand. In the first two countries, more mathematics topics are taught to grade four students than in the latter two countries. In addition, there is a pronounced exposure to the number content area, whereas students in Norway and Sweden are exposed to more topics on data. Finally, the variance is much larger in Norway and Sweden compared to Denmark and Finland. These results point to a clear idea of mathematics as a core school subject, where basic number skills need to be taught extensively during the first years of primary school to provide a solid foundation for later success in Denmark and Finland. In contrast, there is disagreement between teachers and schools in Norway and Sweden about how much mathematics and which topics to teach to grade four students, while favoring the data content domain—which probably due to its association with real-world applications.

4. Similar amounts of mathematics and similar profiles of content exposure do not mean that there are similar effects on outcomes. In Denmark, strong socioeconomic inequalities in content coverage exist. The higher students' SES is, the more content they receive for number and geometry topics, with an additional indirect effect of SES on student achievement in geometry. This means that schooling linearly exacerbates the SES effects of students' family background. The results for Finland in contrast do not point to increased linear differences in content coverage between classes with higher and lower SES students. Thus, there is no additional indirect effect of SES due to this feature of schooling (see footnote 1 for one potential explanation). If the increase is not linear from low- to high-socioeconomic students, traditional linear modelling cannot estimate an effect. Non-linear modelling would be more appropriate.

---

[3] Note that this does not necessarily mean that schooling exacerbates socioeconomic inequalities in mathematics achievement. Imagine for example a scenario where high-SES students are concentrated in selective classes with an emphasis on language, arts, or sports where they receive *fewer* mathematics lessons. This only would reduce socioeconomic inequalities. Imagine in addition that students in special needs classes receive a lot of individualized support adjusted to their needs in learning mathematics.

5. The data for Norway and Sweden again reveal different results. Most of the relations between SES and content exposure are not statistically significant. This can be because of two reasons: the effects are not strong enough, or there may also be non-linear relationships with more content at the lower and upper ends of the SES specter. The descriptive results point to this hypothesis. Moreover, Norwegian students from lower socioeconomic backgrounds are significantly more often exposed to data topics than students from higher SES backgrounds. A similar tendency exists in Sweden for number topics. Such results may point to compensatory approaches by schools and teachers to provide additional opportunities for disadvantaged students in terms of data.

These results and their interpretations reveal that the seriousness of the challenges regarding content coverage varies in the Nordic countries and by content domain. This is particularly interesting given such differences should not exist in education systems that offer comprehensive schools and/or compensate for family disadvantages. An obvious mechanism that plays here may be school fragmentation. Fragmentation could mean differences in continuous family support or in teaching quality so that students learn more or less of the content.

8.4.1 Limitations

One potential limitation of this study is that the opportunity to learn (OTL) may not be directly related to achievement for several reasons. Students could learn content elsewhere, teachers may not accurately report content coverage, and the items may be phrased in such a way that teachers fail to recognize their correct meaning.

Content coverage was assessed via teacher reports, which may be prone to bias in terms of differential benchmarks and accuracy. Teachers may have forgotten or misremembered the content they taught, potentially skewing the results. Daus and Braeken (2018) note that teachers may interpret the term "covered" differently depending on the level of detail they associate with it. Additionally, not all topics are taught by one teacher in every country, which may affect the results. This variability in interpretation can introduce inconsistencies in the data and make it difficult to draw firm conclusions about the relationship between OTL and achievement.

While the data in this study is correlational, Schmidt et al. (2015) argue that in mathematics, OTL and achievement are cumulative by nature, allowing for causal interpretations. Due to the hierarchical structure of mathematics, it is unlikely that more advanced topics are taught before basic ones. This means that even though the data is correlational, the cumulative nature of mathematics learning allows for some level of confidence in interpreting the results as causal relationships.

Additionally, the quality of the OTL might not be sufficient to foster student learning. Factors such as classroom environment, teaching materials, and access to resources can all impact the quality of the OTL. In this context, teaching quality could also serve as a mediating factor between OTL and student achievement.

Finally, the concept of "socioeconomic background" is limited to cultural capital, which can be argued as the most proximal socioeconomic indicator to student learning. However, this limitation should be acknowledged, as other aspects of socioeconomic background, such as financial resources or parental education, may also influence student learning outcomes. By focusing solely on cultural capital, the study may not capture the full extent of socioeconomic effects on student achievement.

8.5 Conclusions and Further Research

The results point to differential effects on various subgroups. However, the mediation model used in this study estimates linear effects, which should be further investigated, especially given that other descriptive evidence supports the presence of a non-linear relationship between content coverage and achievement. Schmidt et al. (2013) also observed that in many countries, achievement tends to decline at the highest levels of content coverage.

Gaining a deeper understanding of the relationship between societal inequalities and educational success or failure is crucial. In the Swedish case, the findings suggest that schools may be successful in countering segregation, a conclusion supported by existing literature. However, it is almost impossible to state this based on solely cross-sectional data.

The Finnish case deserves more in-depth examination. It is essential to explore the potential consequences of excluding special classes to the observed variance in the data. More importantly, researchers should investigate whether students in these special classes would perform better academically if they were integrated into regular classes. Addressing this question is crucial from a policy standpoint.

To build on these findings, future research should focus on examining the non-linear relationship between content coverage and achievement, as well as identifying potential mediating factors that could help explain the varying effects observed across different student subgroups. Longitudinal studies could offer a more comprehensive understanding of the complex interplay between content coverage, societal inequalities, educational policies, and student outcomes over time. By pinpointing the key drivers of educational success and failure, policymakers can develop more effective strategies to ensure equal opportunities for all students, regardless of their socioeconomic background. Ultimately, this will help create a more inclusive and equitable educational environment for future generations.

References

Ahonen, A. K. (2021). Finland: Success through equity—The trajectories in PISA performance. In: N. Crato (Ed.), Improving a country's education. Springer. https://doi.org/10.1007/978-3-030-59031-4_6

Alia, A., Japelj Pavešić, B., & Rožman, M. (2022). Opportunity to learn mathematics and science. In B. Japelj Pavešić, P. Koršňáková, & S. Meinck (Eds.), Dinaric perspectives on TIMSS 2019: Teaching and learning mathematics and science in South-Eastern Europe (pp. 39–64). Springer. https://doi.org/10.1007/978-3-030-85802-5_3

Arnesen, A.-L., & Lundahl, L. (2006). Still social and democratic? Inclusive education policies in the Nordic welfare states. Scandinavian Journal of Educational Research, 50(3), 285–300. https://doi.org/10.1080/00313830600743316

Blossing, U., Imsen, G., & Moos, L. (2014). Nordic schools in a time of change. In U. Blossing, G. Imsen, & L. Moos (Eds.), The Nordic education model (pp. 1–14). Springer. https://doi.org/10.33112/nm.11.1.36

Bostad, I., & Solberg, M. (2022). Rooms of togetherness: Nordic ideals of knowledge in education. In The Nordic education model in context (pp. 125–142). Routledge.

Bourdieu, P. (1986) The forms of capital. In J. Richardson (Ed.), Handbook of theory and research for the sociology of education. Greenwood.

Brese, F. & Mirazchiyski, P. (2013). Measuring students' family background in large-scale international education studies (Issues and methodologies in large-scale assessments: Special issue 2). IERI Monograph Series.

Buchholtz, N., Stuart, A., & Frønes, T. S. (2020). Equity, equality and diversity—Putting educational justice in the Nordic model to a test. In T. S. Frønes et al. (eds.), Equity, equality and diversity in the Nordic model of education, (pp. 13–41). https://doi.org/10.1007/978-3-030-61648-9_2.

Cook, C. J. (1997). Cultural practices and socioeconomic attainment: The Australian experience. Greenwood Press.

Fuchs, T., & Woessman, L. (2007). What accounts for international differences in student performance? A re-examination using PISA data. Empirical Economics, 32, 433–464. https://doi.org/10.1007/s00181-006-0087-0

Guo, C., & Min, W. (2006). The effect of familial economic and cultural capital on educational attainment in China. Journal of Higher Education, 27(24–31).

Daus, S., & Braeken, J. The sensitivity of TIMSS country rankings in science achievement to differences in opportunity to learn at classroom level. Large-scale Assessments in Education, 6(1). https://doi.org/10.1186/s40536-018-0054-1

Hanley, J. E., & McKeever, M. (1997). The persistence of educational inequalities in state-socialist Hungary: Trajectory-maintenance versus counter selection. Sociology of Education, 70(1), 1–18. https://doi.org/10.2307/2673189

Heppt, B., Olczyk, M., & Volodina, A. (2022). Number of books at home as an indicator of socioeconomic status: Examining its extensions and their incremental validity for academic achievement. Social Psychology of Education, 25, 903–928. https://doi.org/10.1007/s11218-022-09704-8

Husén, T. (Ed.). (1967). International study of achievement in mathematics: A comparison of twelve countries (Vol. I). Wiley.

Imsen, G., & Volckmar, N. (2013). The Norwegian school for all: Historical emergence and neoliberal confrontation. In The Nordic education model: 'A school for all encounters neo-liberal policy (pp. 35–55). Springer.

In Matti, T. (Ed.), Northern lights on PISA 2006: Differences and similarities in the Nordic countries. Nordic Council of Ministers. https://www.norden.org/en/publication/northern-lights-pisa-2006

Kosunen, S., & Seppänen, P. (2015). The transmission of capital and a feel for the game: Upper-class school choice in Finland. Acta Sociologica, 58(4), 329–342. https://doi.org/10.1177/0001699315607968

Laukaityte, I., & Wiberg, M. (2017). Using plausible values in secondary analysis in large-scale assessments. Communications in statistics-Theory and Methods, 46(22), 11341–11357.

Lavonen, J., Lie, S., Macdonald, A., Oscarsson, M., Reistrup, C., & Sørensen, H. (2009). *Science education, the science curriculum and PISA 2006* (pp. 31–58).

Luyten, H. (2017). Predictive power of OTL measures in TIMSS and PISA. In J. Scheerens (Ed.), *Opportunity to learn, curriculum alignment and test preparation* (pp. 103–119). Springer. https://doi.org/10.1007/978-3-319-43110-9_5

Marsh, H. W., Trautwein, U., Lüdtke, O., Köller, O., & Baumert, J. (2005). Academic self-concept, interest, grades, and standardized test scores: Reciprocal effects models of causal ordering. *Child development, 76*(2), 397–416.

Møllegaard, S., & Meier Jæger, M. (2015). The effect of grandparents' economic, cultural, and social capital on grandchildren's educational success. *Research in Social Stratification and Mobility, 42*, 11–19. https://doi.org/10.1016/j.rssm.2015.06.004

Mullis, I. V., & Martin, M. O. (2017). *TIMSS 2019 assessment frameworks*. International Association for the Evaluation of Educational Achievement. https://timssandpirls.bc.edu/timss2019/international-results.

Muthen, L. K., & Muthen, B. O. (2020). MPlus (Version 8.5) [Computer software]. Muthen & Muthen. https://www.statmodel.com/index.shtml.

Mullis, I. V. S., Martin, M. O., Foy, P., Kelly, D. L., & Fishbein, B. (2020). *TIMSS 2019 international results in mathematics and science*. TIMSS & PIRLS International Study Center, Boston College. https://timssandpirls.bc.edu/timss2019/international-results/

Opheim, V. (2004). Equity in education: Country analytical report Norway.

Reimer, D., Skovgaard Jensen, S. & Kjeldsen, Ch. (2018). Social inequality in student performance in the Nordic countries: A comparison of methodological approaches. In Reimer, D., Sortkear, B., Oskarsson, M., Nilsen, T., Rasmusson, M., & Nissinen, K. (Eds.), *Northern lights on TIMSS and PISA 2018* (pp. 31–60). Nordic Council of Ministers. https://doi.org/10.6027/TN2018-524

Rutkowski, L., Gonzalez, E., Joncas, M., & Von Davier, M. (2010). International large-scale assessment data: Issues in secondary analysis and reporting. *Educational researcher, 39*(2), 142–151.

Rolfe, V., Strietholt, R., & Yang Hansen, K. (2021). Does inequality in opportunity perpetuate inequality in outcomes? International evidence from four TIMSS cycles. *Studies in Educational Evaluation, 71*. https://doi.org/10.1016/j.stueduc.2021.101086

Sandsør, A. M. J., Zachrisson, H. D., Karoly, L. A., & Dearing, E. (2021, September 13). Achievement gaps by parental income and education using population-level data from Norway. https://doi.org/10.35542/osf.io/unvcy

Scheerens, J. (2017). Opportunity to learn, curriculum alignment and test preparation: A research review. Springer. https://doi.org/10.1007/978-3-319-43110-9

Schmidt, W. H., & McKnight, C. C. (2012). *Inequality for all: The challenge of unequal opportunity in American schools*. Teachers College.

Schmidt, W. H., Zoido, P., & Cogan, L. S. (2013). Schooling matters: Opportunity to learn in PISA 2012 (OECD Education Working Papers No. 95). Organisation for Economic Co-operation and Development (OECD). https://doi.org/10.1787/19939019

Schmidt, W. H., Burroughs, N. A., Zoido, P., & Houang, R. T. (2015). The role of schooling in perpetuating educational inequality: An international perspective. *Educational Researcher, 44*(7), 371–386. https://doi.org/10.3102/0013189X15603982

Sundqvist, C., & Hannås, B. M. (2021). Same vision—different approaches? Special needs education in light of inclusion in Finland and Norway. *European Journal of Special Needs Education, 36*(5), 686–699. https://doi.org/10.1080/08856257.2020.1786911

Trumberg, A., Arneback, E., Bergh, A., & Jämte, J. (2022). Struggling to counter school segregation: A typology of local initiatives in Sweden. *Scandinavian Journal of Educational Research*. https://doi.org/10.1080/00313831.2022.2127877

Vainikainen, M. P., & Harju-Luukkainen, H. (2020). Educational assessment in Finland. *Monitoring student achievement in the 21st century: European policy perspectives and assessment strategies* (pp. 131–142). Springer. https://doi.org/10.1007/978-3-030-38969-7

Wiberg, M., & Rolfsman, E. (2021). Students' Self-reported background SES measures in TIMSS in relation to register SES measures when analysing students' achievements in Sweden. *Scandinavian Journal of Educational Research.* https://doi.org/10.1080/00313831.2021.198 3863

Yang Hansen, K., & Gustafsson, J.-E. (2016). Causes of educational segregation in Sweden—school choice or residential segregation. *Educational Research and Evaluation, 22*(1–2), 23–44. https://doi.org/10.1080/13803611.2016.1178589

Yang Hansen, K., Gustafsson, J.-E., Rosén, M. (2014). School performance differences and policy variations in Finland, Norway and Sweden. In A. B. Kavli & T. Hallvard. (Eds.) *Northern lights on TIMSS and PIRLS 2011: Differences and similarities in the Nordic countries* (pp. 25–48). Nordon: Nordic Council of Ministers. https://www.udir.no/Upload/Forskning/2014/Nlights%20TIMSS%20and%20PIRLS.pdf.

Ye, W., Strietholt, R., & Blömeke, S. (2021). Academic resilience: Underlying norms and validity of definitions. *Educational Assessment, Evaluation and Accountability, 33*(1), 169–202. https://doi.org/10.1007/s11092-020-09351-7.

Sigrid Blömeke (1965–2023) was the Director of the Centre for Educational Measurement in Oslo (CEMO) and Centre for Research on Equality in Education (CREATE). Before moving to Norway, she was a Professor at Humboldt University of Berlin, Germany, and Michigan State University, USA. Blömeke has received Distinguished Research Awards from the University of Oslo, the German Educational Research Association, and the University of Paderborn. She is a member of The Norwegian Academy of Science and Letters and the International Academy of Education.

Chapter 9
Examining the Role of Teaching Quality and Assessment Practice in Reducing Socioeconomic and Ethnic Inequities in Mathematics Achievement

Kajsa Yang Hansen ⓘ**, Victoria Rolfe** ⓘ**, and Nani Teig** ⓘ

9.1 Introduction

Numerous studies have investigated factors that address socioeconomic and ethnic disparities in academic achievement. One area of research, often referred to as opportunity gaps, highlights the importance of teaching quality and practices in promoting educational equity (Akiba, et al., 2007; Klieme et al., 2009; Nilsen & Gustafsson, 2016; Seidel & Shavelson, 2007).

However, a lack of consensus makes defining measures for teaching quality and assessment practices challenging. A widely accepted framework (the three-dimension conceptualization framework) attempting to do so, defines teaching quality as a three-dimensional construct encompassing classroom management, supportive climate, and cognitive activation (Baumert et al., 2010; Creemers & Kyriakides, 2007; Lazarides & Ittel, 2012; Sogunro, 2017). Adopting this framework, this study aims to examine the impact of teaching quality on students' performance across different contexts, thereby identifying effective strategies to reduce educational disparities.

K. Yang Hansen (✉) · V. Rolfe
Department of Education and Special Education, University of Gothenburg, P.O. Box 300, SE 40530 Gothenburg, Sweden
e-mail: kajsa.yang-hansen@ped.gu.se

V. Rolfe
e-mail: victoria.rolfe@gu.se

N. Teig
Department of Teacher Education and School Research, University of Oslo, Postboks 1099, Blindern, 0317 Oslo, Norway
e-mail: nani.teig@ils.uio.no

© The Author(s) 2024
N. Teig et al. (eds.), *Effective and Equitable Teacher Practice in Mathematics and Science Education*, IEA Research for Education 14,
https://doi.org/10.1007/978-3-031-49580-9_9

In recent decades, the education systems in the Nordic countries have evolved towards a high degree of autonomy, privatization, and marketization, leading to increased school segregation in terms of students' intake, educational resources, teacher and teaching quality, and academic achievement (Fjellman et al., 2018; Yang Hansen & Gustafsson, 2016). In this context, this chapter explores the potential influence of teaching and assessment practices on socioeconomic and ethnic equity in mathematics achievement across Nordic classrooms. The study examines how differences in teaching practices, including teaching quality, formative assessment practices, and teachers' emphasis on academic success, contribute to educational equity considering the students' socioeconomic and language backgrounds.

9.2 Review of Previous Research

9.2.1 Students' Socioeconomic and Ethnic Backgrounds and Their Academic Achievement

Socioeconomic background is perhaps the best known and most extensively studied predictor of educational outcomes (see e.g., Sirin, 2005; Strietholt et al., 2019), and is fundamental in determining an individual's life chances (Pinquart & Sörensen, 2000; Tan et al., 2020). As a collective endeavor, schooling is an arena in which students' performance is influenced by their peers. The composition of the student body within a school and a classroom has a profound impact on individual students' achievement, with schools that admit more socioeconomically advantaged students typically displaying stronger academic outcomes (e.g., Agirdag et al., 2012).

Multiple classroom composition effects have been documented as predictors of achievement. Firstly, studying alongside more socioeconomically advantaged peers has been established as a predictor of student outcomes (e.g., van Ewijk & Sleegers, 2010; Yang Hansen et al., 2022). Secondly, increased ethnolinguistic diversity in a classroom is associated with lower average performance in assessments of the national language of the school system (Seuring et al., 2020). Lastly, students benefit from studying alongside high-performing classmates, a phenomenon which particularly benefits high-achieving students (e.g., Lavrijsen et al., 2022).

Domestic studies have observed notable segregation between schools across the Nordic region (e.g., Bernelius & Vilkama, 2019; Rangvid, 2007; Rogne et al., 2021; Yang Hansen & Gustafsson, 2016). These local studies are reflected in pervasive achievement gaps in the Nordic countries across multiple international studies targeting different age groups, including IEA's Trends in International Mathematics and Science Study (TIMSS) 2015 and 2019, and the OECD's Programme for International Student Assessment (PISA) 2015 and 2018 (Mullis et al., 2020; OECD, 2016). Pertinently to this study, composition effects can be observed in the international results for mathematics in TIMSS 2019 grade four, which show variation in achievement scores between schools with differing language use populations and socioeconomic composition (Mullis et al., 2020).

9.2.2 School Emphasis on Academic Success of Teachers and Student's Academic Achievement

While teachers and students are key actors within the classroom, they are not the sole stakeholders in schools as learning communities. Administrators and parents also play an essential role in building the culture within the school. While autonomy to create a school culture varies between countries, one common aspect of school culture that exists across educational systems is the school's emphasis on promoting academic success, which has been shown to relate to student learning (i.e., Bryk & Schneider, 2003; Hoy et al., 2006; Kythreotis et al., 2010; Martin et al., 2013).

In TIMSS, the belief in group success within the school context is conceptualized as the school emphasis on academic success (SEAS). SEAS is compiled from the teacher and principal responses to questionnaire items indicating a supportive environment for academic success (Martin et al., 2013), and a moderate relationship between SEAS and achievement across mathematics, reading, and science has been found among fourth graders. Further, Martin et al.'s (2013) SEAS model has been demonstrated to have high construct validity and be applicable across multiple national contexts. It is also a strong predictor of science achievement in countries like Norway (Nilsen & Gustafsson, 2014). SEAS is strongly associated with classroom SES composition, with teachers reporting higher levels of SEAS in schools with higher SES classrooms (Rolfe et al., 2022). However, while SEAS is highly related to SES, it has been found not to predict achievement when modelling teacher quality and opportunity to learn in Sweden (Rolfe et al., 2022).

9.2.3 Teaching Quality and Student's Academic Achievement

High quality teaching is an established explanatory factor for student achievement (e.g., Blömeke et al., 2016; Darling-Hammond, 2000). However, there is no fixed consensus on what defines teacher quality. Goe (2007) proposes a framework in which various indicators of teacher quality can be grouped as inputs, processes, or outcomes. The present study focuses on the processes and outcomes of teacher quality, particularly on the quality of teaching and assessment practices and student mathematics achievement.

In terms of teaching processes as indicators of teacher quality, the present study examines cognitive activation, classroom management, and supportive climate in line with the three-dimension conceptualization framework of teaching quality (see Baumert et al., 2010; Creemers & Kyriakides, 2007; Lazarides & Ittel, 2012; Sogunro, 2017). Teaching processes involve complex and interweaving sets of behaviors that teachers embody through their professional practice. Many of the findings discussed in the literature which illuminate these processes are the results of locally and regionally administered survey studies (see Goe, 2007).

Cognitive activation, as summarized by Lipowsky et al (2009), involves developing conceptual understanding between new and old content, selecting activities which operate on progressively higher cognitive levels, and engaging in quality discourse with students. Cognitive activation is a significant predictor of student achievement (e.g., Baumert et al., 2010; Li et al., 2021; Lipowsky et al, 2009). This relationship may be related to compositional effects, with Le Donné et al. (2016) proposing a stronger predictive relationship in more socioeconomically advantaged schools.

The previously discussed concept of SEAS is distinct from the concept of classroom climate. SEAS focuses on the prioritization of learning and achievement by multiple stakeholders, including students, teachers, parents, and school leadership—all of whom contribute to the school having a good climate for success (e.g., Nilsen & Gustafsson, 2014). In contrast, a supportive climate for learning can be summarized as allowing individual students exposure to ideas and feedback, supporting them in self-reflection, and promoting behavior modification to facilitate learning (Gibb, 1958).

Established research on the interrelationships between school climate, SES, and achievement is somewhat contradictory (Berkowitz et al., 2017). A positive supportive climate has been suggested as a moderator of the SES-achievement relationship (see, e.g., Cheema & Kitsantas, 2014). School climate is, in turn, seen as a compensatory mechanism for low SES (Brand et al., 2003; Schagen & Hutchison, 2003), or as a phenomenon influenced by the SES of the student body. For example, schools with lower SES students experiencing higher out-of-school social risks may struggle to establish the type of supportive climate, which can affect student outcomes (McCoy et al., 2013).

Additionally, it is worth noting that there is some disagreement in the literature regarding the presence of ethnic bias in teacher ratings of student behavior (see Mason et al., 2014). However, as much of the research in this field is situated in the American context and uses race as its conceptualization of ethnicity, the findings of these studies may not be directly applicable to the Nordic educational context.

9.2.4 Assessment Practices and Students' Academic Achievement

Assessment plays a crucial role in classroom practices, as it informs teachers about students' progress and can be used as a tool to motivate students' learning (e.g., Broadfoot et al., 2002; Gronlund, 2006). Extensive research has explored the impact of teacher assessment practices on student behaviors and outcomes, including learning depth, self-motivation, and achievement (Crooks, 1988). Notably, frequent use of tests in the classroom has a moderate effect size on student attainment (Bangert-Drowns et al., 1991; Yang et al., 2021). However, teachers' perceptions of student achievement may be clouded by implicit biases, such as student ethnicity and socioeconomic background (Darling-Hammond, 1995) or gender (e.g., Guez et al., 2020).

Brookhart (1997) introduced the classroom assessment model, which views the assessment environment as a communal experience for students. In this model, teachers define assessment purposes, set tasks and criteria, provide feedback, and monitor outcomes (see also Brookhart, 2001). The classroom assessment environment influences cognitive and non-cognitive outcomes of students. As students progress through middle school, teachers reported an increased use of assessment tools considered more informative for grading (Martínez et al., 2009). Research conducted in socioeconomically disadvantaged schools has demonstrated that implementing evidence-based instructional and behavior management strategies leads to improved student mathematics knowledge during the elementary and middle years (Reddy et al., 2020). This highlights the importance of effective instructional practices in fostering student progress.

9.2.5 Opportunity to Learn and Students' Academic Achievement

Opportunity to learn (OTL) is a concept which rests on the assumption that an individual will not perform well on tests covering content they have not been taught (e.g., Eggen et al., 1987). While some scholars emphasize the importance of time dedicated to covering content (e.g., Carroll, 1963), as elaborated on in Chap. 2, OTL in TIMSS is considered as the alignment between the *intended curriculum* (formal curricula documents based on national or regional standards), the *implemented curriculum* (what teachers have taught), and the *attained curriculum* (what students have learned) (see Schmidt et al., 1997). However, as noted in Chap. 4, the mathematics curricula in the Nordic countries are structured in such a way that expected learning is not expressed in grade-specific knowledge, but across multi-year phases, which do not align with the grades assessed by TIMSS. This may potentially impact the alignment between the intended and implemented curricula measured through the OTL construct for the national samples in this study.

The relationship between OTL and student achievement has been extensively studied at both the individual (e.g., Schmidt et al., 2013, 2015) and collective levels (see Scheerens et al., 2007; Seidel & Shavelson, 2007). Existing evidence indicates that OTL is a positive predictor of achievement and a mediator of the SES-achievement relationship (e.g., Schmidt et al., 2015). From secondary analyses of ILSA data, it appeared that the strength of this relationship varies depending on the study (PISA vs TIMSS) and the formulation of the construct (Luyten, 2017; Rolfe et al., 2021; Schmidt et al., 2015; Yang Hansen & Streitholt, 2018) and the subject, being evident in mathematics but not science (Luyten, 2017; Rolfe et al., 2021).

9.3 The Hypothesis Model and Research Questions

In summary, the relationships between teaching quality, assessment practices, content coverage, and students' achievement in mathematics are complex and interrelated. It can be hypothesized that teaching quality and assessment practices influence the delivery of curriculum content, which in turn affects students' achievement in a specific subject. It can also be hypothesized that such a conditional classroom mechanism may help to reduce the achievement inequality among students due to their family socioeconomic and ethnic backgrounds (see Appendix 1 for the hypothesis model).

Applying data from TIMSS 2019, the present study aims to address the following research questions by testing the hypothesized conditional mechanism:

1. What are the differences in students' mathematics achievement, socioeconomic status, and ethnic composition across classrooms?
2. Is there socioeconomic and ethnic inequality in students' mathematics achievement?
3. Do socioeconomic and ethnic inequalities in students' mathematics achievement differ significantly across different classrooms?
4. Do teachers' instructional and assessment practices as well as content coverage impact student's mathematics achievement and socioeconomic and ethnic inequality in their achievement?

Considering the institutional differences in the Nordic education systems, the conditional mechanism may vary. This chapter adopts a comparative perspective in exploring the above-mentioned research questions.

9.4 Method

9.4.1 Samples

The current study used TIMSS 2019 data from four Nordic countries, i.e., Denmark, Finland, Norway, and Sweden. Table 9.1 shows the number of individual students and classrooms in the four samples. In total, there are 15,873 students and 966 classrooms to facilitate the current analyses. Finland holds the largest number of sampled classrooms (316) and students (4730), while Denmark has the lowest numbers with 195 classrooms and 3227 students. Norway and Sweden have rather similar sample

Table 9.1 The number of cases in each Nordic education system from TIMSS 2019

Country	Number of individuals	Number of classrooms
Denmark	3227	195
Finland	4730	316
Norway	3951[a]	231
Sweden	3965	224
Total	15,873	966

Note [a] Norwegian sample is from grade five and all other countries' samples are from grade four

sizes. It should be noted that Norway participated in TIMSS 2019 with grade five students and for the rest of the analyzed Nordic countries, grade four students were included in the samples.

9.4.2 Variables

Appendix 2 presents a comprehensive list of all the variables that were involved in the current study. Student socioeconomic status is measured by the number of books at home, ranging from 0 (none to 10 books) to 4 (more than 200 books). The dummy-coded ethnic background is based on student responses as to how often they speak the language of the test at home. If students responded with "always or almost always speak the language of the test at home", they are classified as native, whereas if they reported to "sometimes or never speaking the language of the test at home", they are classified as immigrant. The aggregated cluster mean of SES and ethnicity are used as indicators of socioeconomic and ethnic compositions of classrooms. Teachers' instructional quality (i.e., cognitive activation and supportive climate), assessment practices, their emphasis on academic success (SEAS), and OTL in terms of the percentage of content coverage for the three-mathematics content domains in TIMSS were measured using the teacher questionnaire data (see Appendix 2 for detailed information on these teacher-related constructs). Finally, students' mathematics achievement was captured using five plausible values of the test score and used as the outcome variable in the analysis (see Chap. 3 Analytical Framework for further details).

9.4.3 Analytical Method and Process

Data in TIMSS 2019 were collected using a stratified two-stage cluster sampling design, resulting in a hierarchical data structure where students were nested within classrooms and schools (LaRoche et al., 2020). A two-level structural equation modelling technique (Hox et al., 2017) was therefore required to decompose the total variance of an outcome into individual-level and classroom-level variance components, allowing for accurate standard error estimation based on the correct sources of variation and avoiding type I error in statistical inference. The Intraclass Correlation Coefficient (ICC) quantifies the proportion of total variance in an outcome variable that can be ascribed to variations among students belonging to different classrooms. It serves as a measure of the extent of heterogeneity among students across classrooms, with a higher ICC indicating greater disparities in student outcomes across different classrooms.

Table 9.2 provides information on the ICCs of SES, ethnicity, and mathematics achievement in the four Nordic countries. Sweden held the highest between-classroom differences in mathematics (0.18), SES (0.17), and ethnicity (0.24), indicating a higher level of classroom segregation. Finland also has a relatively high proportion of cross-classroom differences in mathematics scores (0.17) and ethnicity (0.19), but low SES differences. Denmark has the most homogenous classrooms, with the lowest ICCs in mathematics (0.10), SES and ethnicity (both 0.08). The ICCs in Norway were at the intermediate level among the four Nordic countries. In general, cross-classroom differences in the Nordic countries are low when compared internationally, especially in Denmark. However, from an educational inequality perspective, these differences are still substantial and need to be further explained by the teacher-level relevant factors.

The analyses were conducted in multiple steps. Firstly, the sub-dimensions of teaching quality (i.e., cognitive activation and supportive climate) and SEAS were tested for measurement invariance to ensure the comparability of the constructs across the Nordic countries. Metric invariance level was successfully achieved, assuring the comparison of the relationship among the constructs across the four Nordic countries (see Appendix 3). Secondly, a principal component factor score was estimated for these three constructs and used as manifest variables in the two-level models. The average values for formative assessment practices in mathematics classrooms and mathematics OTL were computed, based on the corresponding indicators of the constructs. Subsequently, a series of two-level models at the student and classroom levels were estimated in each of the Nordic countries.

To address research questions 1 and 2, model 1 tested the relationship between SES and ethnicity with mathematics achievement in a two-level structural model for each country, where socioeconomic and ethnic inequality in achievement and classroom segregation can be estimated. Model 2 comprised a set of two-level random slope models, examining whether the socioeconomic and ethnic inequality in mathematics

Table 9.2 Intraclass correlation coefficients of mathematics score, socioeconomic status, and ethnic background

	Denmark	Finland	Norway	Sweden
Mathematics achievement	0.10	0.17	0.12	0.18
Ethnic background	0.08	0.19	0.17	0.24
Socioeconomic status	0.08	0.08	0.11	0.17

achievement varied significantly across different classrooms in the Nordic countries. The result from model 2 provides the answer to research question 3. Finally, model 3 examined whether teaching and assessment practices, teacher emphasis on academic success, and OTL account for the variation in mathematics achievement and socioeconomic and ethnic inequality in achievement (i.e., the random slopes), which will unveil the results of the research question 4 (see Appendix 1).

Modelling was done in Mplus (Muthén & Muthén, 1998–2017) with MLR (maximum likelihood estimator with robust mean and variance) for model 1, where missing data was handled using the expectation–maximization algorithm. The Bayesian estimator was applied for models 2 and 3 with random slopes. For all models, five plausible values for mathematics achievement were used.

9.5 Results

9.5.1 Socioeconomic and Ethnic Inequality in Mathematics Achievement at Student and Classroom Levels

Socioeconomic and ethnic inequality in mathematics achievement is measured by the relationship between students' SES and respective ethnic backgrounds with their mathematics achievement. Table 9.3 shows that at the individual level, the relationship between SES and mathematics was positive and significant in all Nordic countries, and the beta coefficients were rather even, ranging from 0.24 in Denmark and 0.28 in Norway and Sweden. The same is true for the relationship between students' ethnic backgrounds and mathematics achievement, indicating that native students generally have higher achievement than students with migration backgrounds. However, the regression coefficients were rather small after controlling for SES.

At the classroom level, the relationship between classroom SES composition and average mathematics achievement was much higher, compared to those at the individual level. The beta coefficient ranges from around to above 0.70. This implies that about or over half of the variation in mathematics achievement across classrooms can be explained by the differences in the SES composition of the student intake. Finland had the highest SES contextual relationship at 0.79, while Norway had the lowest at 0.66. It is important to note that the ethnic contextual relationship with average mathematics achievement was not significant in Finland and Norway. While in Denmark and Sweden, an additional ethnic contextual relationship was found, being 0.25 and 0.22, respectively. In total, the SES and ethnic context of classrooms explained 62% of the mathematics achievement variation in Finland, followed by 58 percent in Denmark, 52 percent in Sweden and 44% in Norway.

Table 9.3 Relationship between socioeconomic status, ethnic background, and mathematics achievement at individual and classroom levels

Parameters	Denmark		Finland		Norway		Sweden	
	Beta	z	Beta	z	Beta	z	Beta	z
Within-level								
Relationship between mathematics and SES	0.24	12.55	0.26	16.63	0.28	16.55	0.28	19.06
Relationship between mathematics and ethnicity[a]	0.13	5.89	0.08	4.88	0.06	3.34	0.08	4.62
Correlation between SES-ethnicity	0.19	8.90	0.16	8.33	0.21	10.66	0.21	11.54
Between-level								
Relationship between class average math and SES	0.72	8.81	0.79	11.38	0.66	7.35	0.69	8.97
Relationship between class average math and ethnicity	0.25	2.36	ns	1.51	ns	1.13	0.22	2.55
Correlation between class average SES-ethnicity	ns	− 0.25	0.43	4.66	0.49	5.60	0.57	8.27

Note [a] The ethnicity is coded as 1 for native students and 0 for students with a migration background. ns = not statistically significant at $p < 0.05$ level; Beta = Standardized regression coefficients; z = z-value

9.5.2 Testing Random Slopes

The socioeconomic and ethnic inequalities in mathematics achievement among students in the current analysis were captured by the regression coefficients (i.e., the slopes) of mathematics achievement on SES or ethnicity. Previous research indicates that the prevalence of neoliberal ideology has resulted in a global trend towards market-like school systems that emphasize school choice and education provision on demand (e.g., Blossing et al., 2014). Unfortunately, this trend has led to greater social and ethnic segregation as well as quality differences between schools in many countries (e.g., Bonal & Bellei, 2018; OECD, 2012). It can be assumed that some schools and classrooms can effectively compensate for students' disadvantages in sociodemographic backgrounds and help to reduce the SES- or ethnicity-achievement

Table 9.4 Estimated mean and variance of the random slopes

	Denmark		Finland		Norway		Sweden	
	Est.	P. SD	Est.	P. SD	Est.	P. SD	Est.	P. SD
Mean S1	0.169	0.012	0.195	0.011	0.190	0.011	0.182	0.010
Mean S2	0.217	0.033	0.205	0.033	0.116	0.028	0.152	0.026
Variance S1	0.002	0.001	0.002	0.001	0.002	0.001	0.001	0.000
Variance S2	0.021	0.010	0.005	0.005	0.013	0.009	0.003	0.002

Note S1 = relationship between SES and mathematics achievement; S2 = relationship between ethnicity and mathematics achievement; P. SD = Posterior Standard Deviation. Since the estimates for the random slopes were very small, they were presented with three decimals

relationship, while others may fail to fulfil their compensatory mission for disadvantaged students. This mechanism is reflected by the so-called random slopes, meaning that the two slopes vary depending on which school or classroom the student attends.

Table 9.4 shows the estimated mean and variance of the two random slopes. The estimated mean of the slopes S1 (SES-mathematics relationship) and S2 (ethnicity-mathematics relationship) were found to be positive and significant for all Nordic countries. Albeit small in effect, the impact of students' family SES and ethnic background on their mathematics achievement once again was confirmed (e.g., Rolfe & Yang Hansen, 2021). The variance of the two slopes was also statistically significant, implying that the impact of SES and ethnicity of children on their mathematics achievement varies significantly across different classrooms. It should be interesting to explore the classroom-level factors that may be important in accounting for the variation.

9.5.3 Impacts of Teacher-Related Factors on Mathematics Achievement and Inequalities

Two-level path analysis with cross-level interaction and random slopes was conducted to test the hypothesis model (see Sect. 9.3). In that, teaching quality and assessment practices are allowed to affect the delivery of the contents in mathematics, in turn, affect mathematics achievement and SES-achievement and ethnicity-achievement relationships at the individual level. This mechanism was also conditional by teachers' emphasis on academic success, classroom SES and ethnic composition. This is a saturated model with perfect model fit. Figures 9.1, 9.2, 9.3 and 9.4 present the model results of the interrelationships. Note that the non-significant paths are not included in the figures.

Denmark

Based on the results depicted in Fig. 9.1, it is apparent that the teaching quality and teacher assessment practices did not demonstrate any significant relationship

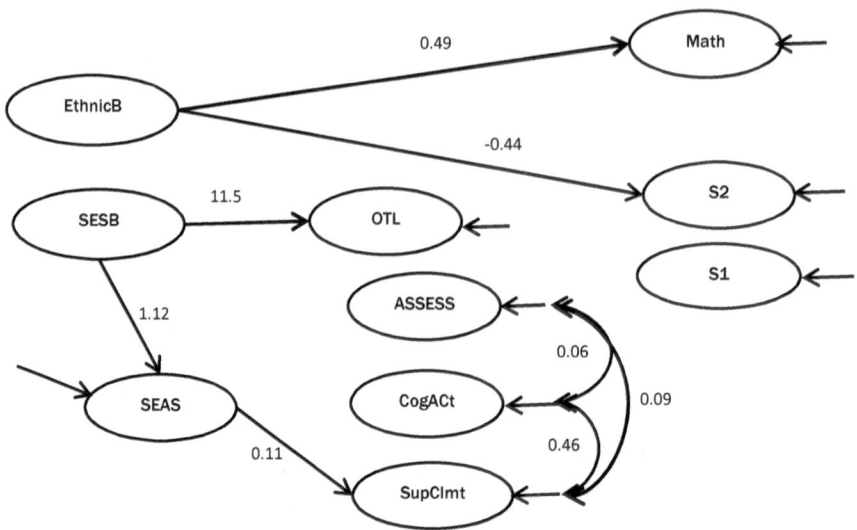

Fig. 9.1 [1]Classroom-level results in Denmark

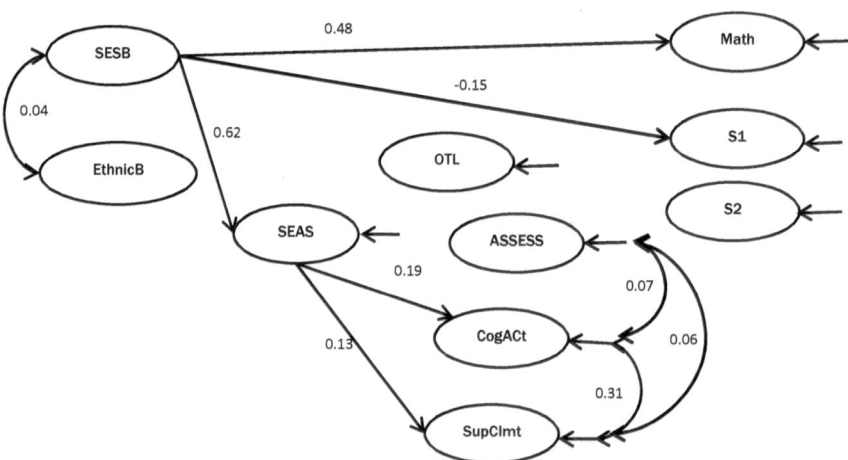

Fig. 9.2 Classroom-level results in Finland

---

[1] For all the figures in the chapter, the variable abbreviations are denoted as following: SESB = class-level socioeconomic composition; EthnicB = class-level ethnic composition; SEAS = teacher's emphasis on student's academic success; CogAct = Cognitive activation practices; ASSESS = assessment practices; SupClim = Supportive climate in the classroom; OTL = content coverage reflecting opportunity to learn; Math = classroom mathematics achievement; S1= random slope between student's socioeconomic status and their mathematics achievement; S2 = random slope between student's ethnic background and mathematics achievement.

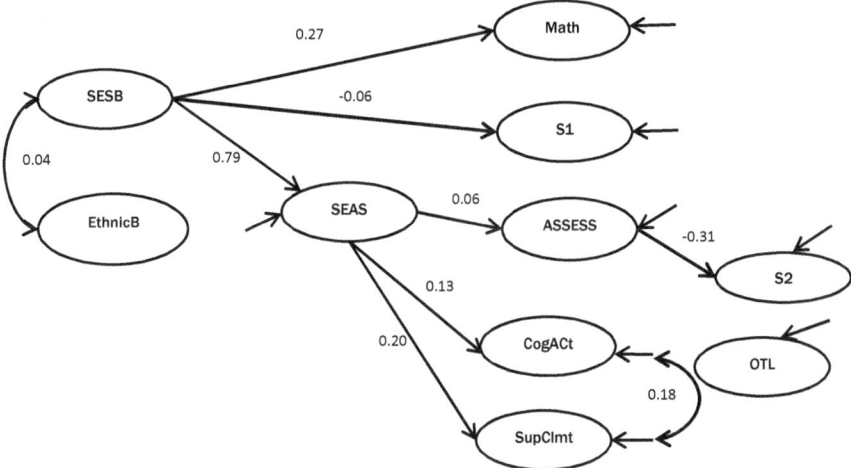

Fig. 9.3 Classroom-level results in Norway

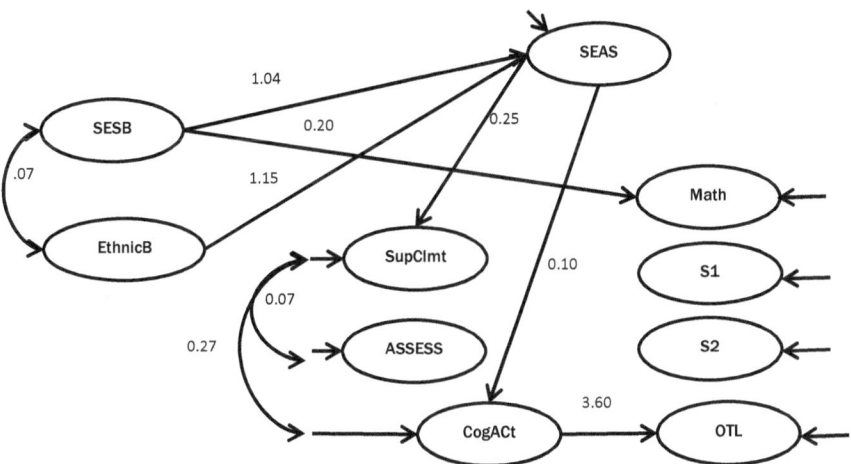

Fig. 9.4 Classroom-level results in Sweden

with the coverage of content in the TIMSS mathematics content domains, nor with mathematics achievement or random slopes. However, contextual factors within the classroom, such as the teacher's emphasis on academic success and the ethnic composition of the classroom, did yield significant impacts. In particular, the ethnic composition of the classroom was positively correlated with the average mathematics score of the classroom (0.49) and served to offset the relationship between ethnicity and individual-level achievement. This suggests that attending a classroom with a greater number of native students mitigates the impact of one's own ethnic background on mathematics achievement (− 0.44), holding other classroom practices and conditions constant. Moreover, the socioeconomic composition of the classroom demonstrated a significant influence on the coverage of content in mathematics domains (11.5) as well as on the teacher's emphasis on academic success (1.12), which in turn influenced the supportive climate of the classroom (0.11).

Finland

Similar to Denmark, there was no significant association between teaching quality and assessment practices and the coverage of content in the content domains of mathematics, nor with mathematics achievement or random slopes in Finland. However, the classroom SES composition had a significant impact on the average mathematics score of the classroom (0.48) and mitigated the relationship between students' family SES and their mathematics achievement (-0.15). In addition, SES composition was significantly linked to teachers' emphasis on academic success (0.62), which, in turn, influenced cognitive activation (0.19) and the supportive climate (0.13) in the classroom.

Norway

In Norway, a similar pattern of relationship was observed between classroom SES composition and average mathematics achievement (0.48), as well as SES-mathematics random slope (-0.15). This suggests that a higher SES composition not only has an impact on classroom mathematics achievement but also reduces the influence of students' family SES on their mathematics achievement. Additionally, classroom SES composition had an indirect effect on the ethnic-mathematics random slope through teachers' emphasis on academic success (0.79) and assessment practices (-0.31). High SES composition classrooms reduce the impact of ethnic composition on students' mathematics achievement through teachers' emphasis on academic success and assessment practices (-0.26). Furthermore, classroom SES composition indirectly affected cognitive activation (0.13) and supportive climate (0.20) through teachers' emphasis on academic success.

Sweden

Both ethnic and SES classroom compositions had an impact on teachers' emphasis on academic success (1.04 and 1.15, respectively), which in turn, had a positive relationship with supportive climate (0.25), cognitive activation (0.10), and content coverage (3.60). No significant relationship was found between any teacher or classroom-level variables and the two random slopes. Furthermore, the SES composition had a positive effect on the average mathematics achievement of the class (0.20).

9.6 Discussion and Conclusions

The global prevalence of neoliberal ideology, especially market mechanisms, such as school choice, autonomy, and competition, in recent decades has significantly transformed the unified and egalitarian Nordic model into systems with increasingly diverse educational practices across systems and intensified segregation along socioeconomic and ethnic lines (e.g., Blossing et al., 2014; Yang Hansen & Gustafsson, 2019).

Using the fourth grade TIMSS 2019 data from Denmark, Finland, Norway, and Sweden, this chapter aimed to identify similarities and disparities in the impact of teaching quality (i.e., cognitive activation and supportive climate), assessment practices, and content coverage on socioeconomic and ethnic inequalities in mathematics achievement across the Nordic countries. The study also analyzed how these relationships are influenced by the sociodemographic context of the classroom, including classroom SES and ethnic compositions, and the teacher's emphasis on academic success.

It was revealed that the socioeconomic and ethnic contexts of the classroom played important roles in students' mathematics achievement in all Nordic countries analyzed. Furthermore, attending a school or classroom with a high proportion of native students and a high socioeconomic status was found to have a compensatory effect in reducing the effect of family socioeconomic status and ethnic background on students' mathematics achievement, therefore, beneficial for students' learning outcomes and social ethnic inequality. Another common feature found in all four Nordic countries studied was that classroom sociodemographic contexts, especially the socioeconomic composition of a classroom, positively associated with teachers' emphasis on students' academic success and, in turn, promoted classroom teaching quality. In Sweden, the ethnic composition seemed to be equally important. This may be attributed to the fact that Sweden has been a leader in extensive changes marked by decentralization and significant marketization and privatization (e.g., Lundahl, 2016). This is also evident in the high disparities in mathematics achievement, SES, and ethnic composition across Swedish classrooms/schools.

Lubienski et al. (2022) highlighted that implementing such policies could lead to social segregation, as families might select schools based on non-academic social factors, and schools could adopt practices that restrict enrollment for less favored students. In Sweden, for example, the universal voucher program with free choice of schools led to increased segregation of native and immigrant students, as well as further stratification based on parental education (e.g., Yang Hansen & Gustafsson, 2016, 2019). In many countries, teachers tend to prefer working in schools with a more socioeconomically advantaged student composition or even a higher proportion of 'white' students (Bonesrønning et al., 2005; Glassow & Jerrim, 2022; Hansson & Gustafsson, 2016). It is evident that schools with a higher socioeconomic status composition and a higher proportion of native students often form a better school ethos for learning and access well-qualified and motivated teachers with a strong emphasis on student's academic success (e.g., Akiba et al., 2007; Han, 2018). The differentiated learning environment, peer groups, and teacher resources contribute to the achievement gap between school and student outcomes.

However, the results regarding the impact of teachers' instructional and assessment practices and content coverage on mathematics achievement were unexpected. None of these factors considered was found to have a significant effect on classroom mathematics achievement or the socioeconomic and ethnic inequality of mathematics achievement, except for Norway. It was demonstrated that mathematics performance differences between native Norwegian children and children with migration backgrounds seem to be reduced by assessment practices. In other words, in a classroom where the teacher applied different types of formative assessment practices more frequently, the mathematics achievement gap between native and non-native students was smaller. The compensatory effect of assessment practices was reinforced by classroom socioeconomic context and teachers' emphasis on students' academic success. In Norway, formative assessment practices are often used in the classrooms to provide feedback to students (Gamlem & Smith, 2013; Havnes et al., 2012). This approach is recognized as a useful tool for supporting student learning and guiding teacher instruction (Havnes et al., 2012). By identifying areas of strength and weakness, teachers can adjust their instruction to better support the learning needs of their students. This approach may offer potential benefits for students with a migration background, as it has the potential to address their unique learning needs.

To conclude, the study has highlighted several areas that require attention in both teacher practices and policy innovation. These efforts may contribute to enhancing educational equity and promoting school desegregation.

Promote socioeconomically and ethnically diverse classrooms

Education policies could promote socioeconomically and ethnically diverse classrooms by ensuring that schools are not segregated based on students' backgrounds. This could be achieved through policies that promote school integration, such as zoning policies that promote diversity in student populations.

Increase emphasis on academic success

Teachers and schools should emphasize academic success and set high expectations for all students, regardless of their socioeconomic or ethnic background. This could include providing additional support for struggling students, setting academic goals, and creating a positive and supportive learning environment.

Improve assessment practices

Education policies and practices should aim to improve assessment practices for learning, including providing additional training for teachers on how to assess student performance fairly and equitably, and ensuring that assessment practices are aligned with curriculum standards and goals. By doing so, education systems can promote a more accurate understanding of student learning and progress, which can help to reduce performance differences between students.

Appendices

Appendix 1 Hypothetical Model of the Conditioning Mechanism Among Teacher Practices and Student's Level and Inequality of Mathematics Achievement

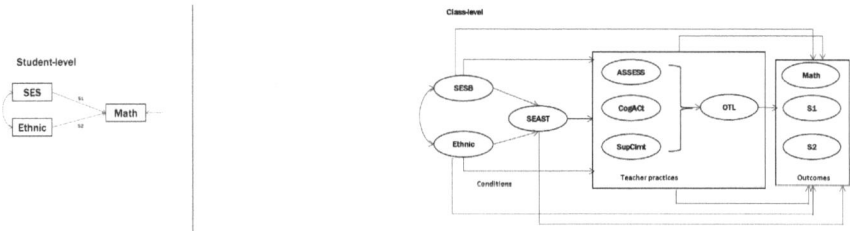

Appendix 2 List of the Variables Used in the Analysis

Construct	Variables	Items in the teacher questionnaire	Scale	Mean (SD) reliability
SES	Books	How many books do you have in your home?	5-scale: 0 = 0–10 1 = 11–25 2 = 26–100 3 = 101–200 4 = > 200	$M_{Den} =$ 1.84 (1.34) $M_{Fin} = 2.11$ (1.07) $M_{Nor} =$ 2.07 (1.15) $M_{Swe} =$ 2.02 (1.22)
Ethnicity	Language	How often do you speak the test language at home?	Dummy coded 0 = Immigrants 1 = Swedish	$M_{Den} =$ 0.74 (0.44) $M_{Fin} = 0.85$ (0.36) $M_{Nor} =$ 0.66 (0.48) $M_{Swe} =$ 0.65 (0.49)

(continued)

(continued)

Construct	Variables	Items in the teacher questionnaire	Scale	Mean (SD) reliability
Teachers' emphasis academic success	**How would you characterise each of the following within your school?**		5-scaled variables from 5 = very high to 1 = very low	α_{den} = 0.852 α_{fin} = 0.849 α_{nor} = 0.848 α_{swe} = 0.848
	ATBG06A	Teachers' understanding of the school's curricular goals		
	ATBG06B	Teachers' degree of success in implementing the school's curriculum		
	ATBG06C	Teachers' expectations for student achievement		
	ATBG06D	Teachers' ability to inspire student		
	ATBG06E	Parental involvement in school activities		
	ATBG06F	Parental commitment to ensure that students are ready to learn		
	ATBG06G	Parental expectations for student achievement		
	ATBG06H	Parental support for student achievement		
	ATBG06I	Students' desire to do well in school		
	ATBG06J	Students' ability to reach school's academic goal		
	ATBG06K	Students' respect for classmates who excel academically		
	ATBG06L	Collaboration between school leadership and teachers to plan instruction		
Supportive climate	**How often do you do the following in teaching this class?**		4-scaled variables from 4 = every or almost every lesson to 1 = never	α_{den} = 0.807 α_{fin} = 0.736 α_{nor} = 0.777 α_{swe} = 0.733
	ATBG12A	Relate the lesson to students' daily lives		
	ATBG12B	Ask students to explain their answers		
	ATBG12C	Bring interesting materials to class		
	ATBG12D	Ask students to complete challenging exercises that require them to go beyond the instruction		

(continued)

(continued)

Construct	Variables	Items in the teacher questionnaire	Scale	Mean (SD) reliability
	ATBG12E	Encourage classroom discussions among students		
	ATBG12F	Link new content to students' prior knowledge		
	ATBG12G	Ask students to decide their own problem-solving procedures		
	ATBG12H	Encourage students to express their ideas in class		
Cognitive activation	**In teaching mathematics to this class, how often do you usually ask students to do the following?**		4-scaled variables from 4 = every or almost every lesson to 1 = never	$\alpha_{den} = 0.698$ $\alpha_{fin} = 0.652$ $\alpha_{nor} = 0.604$ $\alpha_{swe} = 0.686$
	ATBM02A	Listen to me explain new mathematics content		
	ATBM02B	Listen to me explain how to solve problems		
	ATBM02C	Memorize rules, procedures, and facts		
	ATBM02D	Practice procedures on their own		
	ATBM02E	Apply what they have learned to new problem situations on their own		
	ATBM02F	Work problems together in the whole class with direct guidance from me		
	ATBM02G	Work in mixed ability groups		
	ATBM02H	Work in same ability groups		
Opportunity to learn	**In your view, to what extent do the following limit how you teach this class?**		Scale variable	$M_{Den} = 73.2\ (19.0)$ $M_{Fin} = 74.6\ (18.2)$ $M_{Nor} = 73.5\ (16.7)$ $M_{Swe} = 64.1\ (19.2)$
	PTpNum	Pct Students Taught Number Topics		
	PTpGeo	Pct Students Taught Means and Geo Topics		
	PTpData	Pct Std Taught Data Display Topics		

(continued)

(continued)

Construct	Variables	Items in the teacher questionnaire	Scale	Mean (SD) reliability
Mathematics assessment practices	**How much importance do you place on the following assessment strategies in mathematics?**		3-scaled variables, 3 = a lot to 1 = none	M_{Den} = 1.25 (0.30) M_{Fin} = 1.31 (0.25) M_{Nor} = 1.20 (0.27) M_{Swe} = 1.37 (0.31)
	ATBM07A	Observing students as they work		
	ATBM07B	Asking students to answer questions during class		
	ATBM07C	Short, regular written assessments		
	ATBM07D	Longer tests (e.g., unit tests or exams)		
	ATBM07E	Long-term projects		

Note The numbers in parentheses are standard deviation

The bolded text in Appendix 2 indicates that the text is from the Head part of a question (where the variables below are sub-questions) in the TIMSS questionnaire. These sections of text can be unbolded to match the house style

Appendix 3 Results from Measurement Invariance Tests and Homogeneity of Variance Tests of the Constructs in the Current Study

To assess the adequacy of model fit in confirmatory factor analysis (CFA), it is necessary to examine various fit indices, including Root Mean Square Error of Approximation (RMSEA), comparative fit index (CFI), and Standardized Root Mean Square Residual (SRMR). A simulation study examining the rates of rejection for both correctly specified and misspecified models, Hu and Bentler (1999) recommended that an RMSEA value below 0.06 and a CFI value above 0.95, along with an SRMR value below 0.08, are generally indicative of a satisfactory fit.

Based on the fit indices of the measurement models presented in the tables below, the metric invariance model demonstrated favourable model fit across all three constructs. This is further supported by the differences observed in the RMSEA, CFI, and SRMR values between the configural and metric invariance models, aligning with the threshold values recommended by Svetina and Rutkowski (2017).

Model fit indices for measurement invariance test for school emphasis on student academic success (SEAS).

Country	χ^2 (df)	RMSEA	CFI	SRMR
Configural	0(0)	0.000	1.000	0.000
Metric	5.240(6)	0.000	1.000	0.047
Scalar	58.454(12)	0.135	0.872	0.077
	$\Delta \chi^2$ (df)	Δ RMSEA	Δ CFI	Δ SRMR
Configural-metric	5.240(6)[ns]	0.000	0.000	− 0.047
Configural-scalar	58.454(12)	-0.135	0.128	− 0.077

Model fit indices for measurement invariance test for cognitive activation.

Country	χ^2 (df)	RMSEA	CFI	SRMR
Configural	116.306(68)	0.060	0.930	0.053
Metric	150.912(89)	0.059	0.910	0.076
Scalar	433.268(110)	0.122	0.532	0.142
	$\Delta \chi^2$ (df)	Δ RMSEA	Δ CFI	Δ SRMR
Configural-metric	34.675(21)[ns]	0.001	0.020	− 0.023
Configural-scalar	324.112(42)	− 0.062	− 0.532	− 0.89

Model fit indices for measurement invariance test for supportive climate.

Country	χ^2 (df)	RMSEA	CFI	SRMR
Configural	144.242(72)	0.070	0.926	0.050
Metric	169.967(93)	0.063	0.921	0.070
Scalar	290.185(114)	0.087	0.820	0.090
	$\Delta \chi^2$ (df)	Δ RMSEA	Δ CFI	Δ SRMR
Configural-metric	26.387(21)[ns]	0.007	0.005	− 0.020
Configural-scalar	145.482(42)	− 0.017	0.106	− 0.040

Note ns = not significant. Δ denotes the difference in the respective model fit indices between configural and metric or configural and Scalar invariance models

References

Agirdag, O., Van Houtte, M., & Van Avermaet, P. (2012). Why does the ethnic and socio-economic composition of schools influence math achievement? The role of sense of futility and futility culture. *European Sociological Review, 28*(3), 366–378. https://doi.org/10.1093/esr/jcq070

Akiba, M., LeTendre, G. K., & Scribner, J. P. (2007). Teacher quality, opportunity gap, and national achievement in 46 countries. *Educational Researcher, 36*(7), 369–387. https://doi.org/10.3102/0013189X07308739

Bangert-Drowns, R. L., Kulik, J. A., & Kulik, C.-L. C. (1991). Effects of frequent classroom testing. *The Journal of Educational Research, 85*(2), 89–99. http://www.jstor.org/stable/27540459

Baumert, J., Kunter, M., Blum, W., Brunner, M., Voss, T., Jordan, A., Tsai, Y.-M., Klusmann, U, Krauss, S, & Neubrand, M. (2010) Teachers' mathematical knowledge, cognitive activation in the classroom, and student progress. *American Educational Research Journal, 47*(1), 133–180. https://doi.org/10.3102/0002831209345157

Berkowitz, R., Moore, H., Astor, R. A., & Benbenishty, R. (2017). A research synthesis of the associations between socioeconomic background, inequality, school climate, and academic achievement. *Review of Educational Research, 87*(2), 425–469. https://doi.org/10.3102/0034654316669821

Bernelius, V., & Vilkama, K. (2019). Pupils on the move: School catchment area segregation and residential mobility of urban families. *Urban Studies, 56*(15), 3095–3116. https://doi.org/10.1177/0042098019848999

Blossing, U., Imsen, G., & Moos, L. (Eds.) (2014). *The Nordic education model*. Springer. https://doi.org/10.1007/978-94-007-7125-3

Blömeke, S., Olsen, R. V., & Suhl, U. (2016). Relation of student achievement to the quality of their teachers and instructional quality. In: Nilsen, T., Gustafsson, JE. (Eds.), *Teacher quality, instructional quality and student outcomes. IEA research for education* (Vol 2). Springer. https://doi.org/10.1007/978-3-319-41252-8_2

Bonesrønning, H., Falch, T., & Strøm, B. (2005). Teacher sorting, teacher quality, and student composition. *European Economic Review, 49*, 457–483. https://doi.org/10.1016/S0014-2921(03)00052-7

Brand, S., Felner, R., Shim, M., Seitsinger, A., & Dumas, T. (2003). Middle school improvement and reform: Development and validation of a school-level assessment of climate, cultural pluralism, and school safety. *Journal of Educational Psychology, 95*(3), 570. https://doi.org/10.1037/0022-0663.95.3.570

Broadfoot, P. M., Daugherty, R., Gardner, J., Harlen, W., James, M., & Stobart, G. (2002). *Assessment for learning: 10 principles*. University of Cambridge. https://dspace.stir.ac.uk/handle/1893/32458

Brookhart, S. M. (1997). A theoretical framework for the role of classroom assessment in motivating student effort and achievement. *Applied Measurement in Education, 10*, 161–180. https://doi.org/10.1207/s15324818ame1002_4

Brookhart, S. M. (2001). Successful students' Formative and summative uses of assessment information. *Assessment in Education: Principles, Policy & Practice, 8*(2), 153–169. https://doi.org/10.1080/09695940123775

Bonal, X., & Bellei, C. (2018) *Understanding school segregation. Patterns, causes and consequences of spatial inequalities in education*. Bloomsbury.

Bryk, A. S., & Schneider, B. L. (2003). Trust in schools: A core resource for school reform. *Educational Leadership, 60*(6), 40–45.

Carroll, J. B. (1963). A model of school learning. *Teachers College Record: THe Voice of Scholarship in Education, 64*(8), 0000. https://doi.org/10.1177/016146816306400801

Cheema, J. R., & Kitsantas, A. (2014). Influences of disciplinary classroom climate on high school student self-efficacy and mathematics achievement: A look at gender and racial–ethnic differences. *International Journal of Science and Mathematics Education, 12*(5), 1261–1279. https://doi.org/10.1007/s10763-013-9454-4

Creemers, B., & Kyriakides, L. (2007). The dynamics of educational effectiveness: A contribution to policy, practice and theory in contemporary schools: Routledge.

Crooks, T. J. (1988). The impact of classroom evaluation practices on students. *Review of Educational Research, 58*(4), 438–481.

Darling-Hammond, L. (1995). Inequality and access to knowledge. In J. A. Banks (Ed.), *The handbook of research on multicultural education* (pp. 465–483). Macmillan.

Darling-Hammond, L. (2000). Teacher quality and student achievement. *Education Policy Analysis Archives, 8*, 1. https://doi.org/10.14507/epaa.v8n1.2000

Eggen, T. J. H. M., Pelgrum, W. J., & Plomp, T. (1987). The implemented and attained mathematics curriculum: Some results of the second international mathematics study in The Netherlands. *Studies in Educational Evaluation, 13*(1), 119–135. https://doi.org/10.1016/S0191-491 X(87)80026-3

Fjellman, A. M., Yang Hansen, K., & Beach, D. (2018). School choice and implications for equity: The new political geography of the Swedish upper secondary school market. *Educational Review, 71*(4). https://doi.org/10.1080/00131911.2018.1457009

Gamlem, S. M., & Smith, K. (2013). Student perceptions of classroom feedback. *Assessment in Education: Principles, Policy & Practice, 20*(2), 150–169. https://doi.org/10.1080/0969594x. 2012.749212

Gibb, J. R. (1958). A climate for learning. *Adult Education Quarterly, 9*(1), 19–21. https://doi.org/ 10.1177/074171365800900106

Glassow, L. N., & Jerrim, J. (2022). Is inequitable teacher sorting on the rise? Cross-national evidence from 20 years of TIMSS. *Large-Scale Assessments in Education, 10*(1), 6. https://doi. org/10.1186/s40536-022-00125-9

Goe, L. (2007). *The link between teacher quality and student outcomes: A research synthesis.* National Comprehensive Center for Teacher Quality.

Gronlund, N. E. (2006). *Assessment of student achievement* (8th ed.). Pearson.

Guez, A., Peyre, H., & Ramus, F. (2020). Sex differences in academic achievement are modulated by evaluation type. *Learning and Individual Differences, 83–84*, 101935. https://doi.org/10.1016/ j.lindif.2020.101935

Han, S. W. (2018). School-based teacher hiring and achievement inequality: A comparative perspective. *International Journal of Educational Development, 61*, 82–91. https://doi.org/10.1016/j.ije dudev.2017.12.004

Hansson, A., & Gustafsson, J. E. (2016). Pedagogisk Segregation: Lararkompetens i en Svenska Grundskolan ur ett Likvardighetsperspektiv [Pedagogical Segregation: Teacher Competence in Swedish Elementary School from an Equality Perspective, in Swedish]. *Pedagogisk Forskning i Sverige, 21*, 56–78.

Havnes, A., Smith, K., Dysthe, O., & Ludvigsen, K. (2012). Formative assessment and feedback: Making learning visible. *Studies in Educational Evaluation, 38*(1), 21–27. https://doi.org/10. 1016/j.stueduc.2012.04.001

Hox, J. J., Moerbeek, M., & Van de Schoot, R. (2017). *Multilevel analysis: techniques and applications.* Routledge. https://doi.org/10.4324/9781315650982

Hoy, W. K., Tarter, C. J., & Hoy, A. W. (2006). Academic optimism of schools: A force for student achievement. *American Educational Research Journal, 43*(3), 425–446. https://doi.org/10.3102/ 00028312043003425

Hu, L., & Bentler, P. M. (1999). Cutoff criteria for fit indexes in covariance structure analysis: Conventional criteria versus new alternatives. *Structural Equation Modeling: A Multidisciplinary Journal, 6*(1), 1–55. https://doi.org/10.1080/10705519909540118

Klieme, E., Pauli, C., & Reusser, K. (2009). The Pythagoras study: Investigating effects of teaching and learning in Swiss and German mathematics classrooms. In T. Janik & T. Seidel (Eds.), *The power of video studies in investigating teaching and learning in the classroom* (pp. 137–160). Waxmann Publishing Co.

Kythreotis, A., Pashiardis, P., & Kyriakides, L. (2010). The influence of school leadership styles and culture on students' achievement in Cyprus primary schools. *Journal of Educational Administration, 48*, 218–240. https://doi.org/10.1108/09578231011027860

LaRoche, S., Joncas, M., & Foy, P. (2020). Sample design in TIMSS 2019. In M. O. Martin, M., von Davier, & I. V.S. Mullis, (Eds), *Methods and procedures: TIMSS 2019 Technical Report* (pp. 3.1–3.33). TIMSS & PIRLS International Study Center, Boston College. https://timssandp irls.bc.edu/timss2019/methods/chapter-3.html

Lavrijsen, J., Dockx, J., Struyf, E., & Verschueren, K. (2022). Class composition, student achievement, and the role of the learning environment. *Journal of Educational Psychology, 114*(3), 498–512. https://doi.org/10.1037/edu0000709

Lazarides, R., & Ittel, A. (2012). Instructional quality and attitudes toward mathematics: do self-concept and interest differ across students' patterns of perceived instructional quality in mathematics classrooms? *Child Development Research*, 2012. https://doi.org/10.1155/2012/813920

Le Donné, N., Fraser, P., & Bousquet, G. (2016). *Teaching strategies for instructional quality: Insights from the TALIS-PISA link data*. OECD Education Working Papers 148, OECD Publishing.

Li, H., Liu, J., Zhang, D., & Liu, H. (2021). Examining the relationships between cognitive activation, self-efficacy, socioeconomic status, and achievement in mathematics: A multi-level analysis. *British Journal of Educational Psychology, 91*(1), 101–126. https://doi.org/10.1111/bjep.12351

Lipowsky, F., Rakoczy, K., Pauli, C., Drollinger-Vetter, B., Klieme, E., & Reusser, K. (2009). Quality of geometry instruction and its short-term impact on students' understanding of the Pythagorean Theorem. *Learning and Instruction, 19*(6), 527–537. https://doi.org/10.1016/j.learninstruc.2008.11.001

Lubienski, C., Perry, L. B., Kim, J., & Canbolat, Y. (2022). Market models and segregation: Examining mechanisms of student sorting. *Comparative Education, 58*(1), 16–36. https://doi.org/10.1080/03050068.2021.2013043

Lundahl, L. (2016). Equality, inclusion and marketization of Nordic education: Introductory notes. *Research in Comparative and International Education, 11*(1), 3–12.

Luyten, H. (2017). Predictive power of OTL measures in TIMSS and PISA. In J. Scheerens (Ed.), *Opportunity to learn, curriculum alignment and test preparation: A research review* (pp. 103–120). Springer. https://doi.org/10.1007/978-3-319-43110-9

Martin, M. O., Foy, P., Mullis, I. V. S., & O'Dwyer, L. M. (2013). Effective schools in reading, mathematics, and science at fourth grade. In M. O. Martin & I. V. S. Mullis (Eds.), *TIMSS and PIRLS 2011: Relationships among reading, mathematics, and science achievement at the fourth grade—Implications for early learning* (pp. 109–178). TIMSS & PIRLS International Study Center.

Martínez, J., Stecher, B., & Borko, H. (2009). Classroom assessment practices, teacher judgments, and student achievement in mathematics: Evidence from the ECLS. *Educational Assessment, 14*(2), 78–102. https://doi.org/10.1080/10627190903039429

Mason, B. A., Gunersel, A. B., & Ney, E.A. (2014), Cultural and ethnic bias in teacher ratings of behavior: A criterion-focused review. *Psychology in the Schools, 51*, 1017–1030. https://doi.org/10.1002/pits.21800

McCoy, D. C., Roy, A. L., & Sirkman, G. M. (2013). Neighborhood crime and school climate as predictors of elementary school academic quality: A cross-lagged panel analysis. *American Journal of Community Psychology, 52*, 128–140. https://doi.org/10.1007/s10464-013-9583-5

Mullis, I. V. S., Martin, M. O., Foy, P., Kelly, D. L., & Fishbein, B. (2020). *TIMSS 2019 international Results in mathematics and science*. TIMSS & PIRLS International Study Center. Boston College. https://timssandpirls.bc.edu/timss2019/international-results/

Muthén, L. K., & Muthén, B. O. (1998–2017). *Mplus user's guide* (8th ed.). Muthén & Muthén.

Nilsen, T. & Gustafsson, J.-E. (Eds.) (2016). Teacher quality, instructional quality and student outcomes. *IEA research for education* (Vol. 2). https://doi.org/10.1007/978-3-319-41252-8

Nilsen, T., & Gustafsson, J.-E. (2014). School emphasis on academic success: exploring changes in science performance in Norway between 2007 and 2011 employing two-level SEM. *Educational Research & Evaluation, 20*. https://doi.org/10.1080/13803611.2014.941371

OECD. (2012). *Equity and quality in education: Supporting disadvantaged students and schools*. OECD Publishing. https://doi.org/10.1787/9789264130852-en

OECD. (2016). *PISA 2015 results (Vol. I): Excellence and equity in education*. OECD Publishing. https://doi.org/10.1787/9789264266490-en

OECD. (2019). *PISA 2018 results (Vol. II): Where all students can succeed*. OECD Publishing. https://doi.org/10.1787/b5fd1b8f-en

Pinquart, M., & Sörensen, S. (2000). Influences of socioeconomic status, social network, and competence on subjective well-being in later life: A meta-analysis. *Psychology and Aging, 15*(2), 187–224. https://doi.org/10.1037/0882-7974.15.2.187

Rangvid, B. S. (2007). Living and learning separately? Ethnic segregation of school children in Copenhagen. *Urban Studies, 44*(7), 1329–1354. https://doi.org/10.1080/00420980701302338

Reddy, L. A., Lekwa, A., Dudek, C., Kettler, R., & Hua, A. (2020). Evaluation of teacher practices and student achievement in high-poverty schools. *Journal of Psychoeducational Assessment, 38*(7), 816–860.

Rogne, A. F., Borgen, S. T., & Nordrum, E. (2021). School Segregation and native flight. Evidence from school catchment area borders. https://doi.org/10.31235/osf.io/xykdg

Rolfe, V., & Yang Hansen, K. (2021). Family socioeconomic and migration background mitigating educational-relevant inequalities. In: Nilsen, T., Stancel-Piątak, A., Gustafsson, JE. (Eds.), *International handbook of comparative large-scale studies in education. Springer international handbooks of education.* Springer. https://doi.org/10.1007/978-3-030-38298-8_50-1

Rolfe, V., Yang Hansen, K., & Strietholt, R. (2022). Integrating educational quality and educational equality into a model of mathematics performance. *Studies in Educational Evaluation, 74,* 101171. https://doi.org/10.1016/j.stueduc.2022.101171

Rolfe, V., Strietholt, R., & Yang Hansen, K. (2021). Does inequality in opportunity perpetuate inequality in outcomes? International evidence from four TIMSS cycles. *Studies in Educational Evaluation, 71,* 101086. https://doi.org/10.1016/j.stueduc.2021.101086

Schagen, I., & Hutchison, D. (2003). Adding value in educational research—The marriage of data and analytical power. *British Educational Research Journal, 29*(5), 749–765. https://doi.org/10.1080/0141192032000133659

Scheerens, J., Luyten, H., Steen, R., & Luyten-de Thouars, Y. (2007). *Review and meta-analyses of school and teaching effectiveness.* University of Twente.

Schmidt, W. H., Burroughs, N. A., Zoido, P., & Houang, R. T. (2015). The role of schooling in perpetuating educational inequality: An international perspective. *Educational Researcher, 44*(7), 371–386. https://doi.org/10.3102/0013189x15603982

Schmidt, W. H., Raizen, S., Britton, E., Bianchi, L., & Wolfe, R. (1997). *Many visions, many aims volume 2: A cross-national investigation of curricular intentions in school Science.* Springer. https://doi.org/10.1007/978-0-306-47208-4

Schmidt, W. H., Zoido, P., & Cogan, L. (2013). Schooling matters: Opportunity to learn in PISA 2012. In *OECD Education Working Papers.* https://doi.org/10.1787/19939019

Seidel, T., & Shavelson, R. J. (2007). Teaching effectiveness research in the past decade: The role of theory and research design in disentangling meta-analysis results. *Review of Educational Research, 77*(4), 454–499. https://doi.org/10.3102/0034654307310317

Seuring, J., Rjosk, C., & Stanat, P. (2020). Ethnic classroom composition and minority language use among classmates: Do peers matter for students' language achievement? *European Sociological Review, 36*(6), 920–936. https://doi.org/10.1093/esr/jcaa022

Sirin, S. R. (2005). Socioeconomic status and academic achievement: A meta-analytic review of research. *Review of Educational Research, 75*(3), 417–453. https://doi.org/10.3102/00346543075003417

Sogunro, O. A. (2017). Quality instruction as a motivating factor in higher education. *International Journal of Higher Education, 6*(4), 173–184. https://doi.org/10.5430/ijhe.v6n4p173

Strietholt, R., Gustafsson, J.-E., Hogrebe, N., Rolfe, V., Rosén, M., Steinmann, I., & Yang Hansen, K. (2019). The impact of education policies on socioeconomic inequality in student achievement: A review of comparative studies. In L. Volante, S. V. Schnepf, J. Jerrim, & D. A. Klinger (Eds.), *Socioeconomic inequality and student outcomes: Cross-national trends, policies, and practices* (pp. 17–38). Springer. https://doi.org/10.1007/978-981-13-9863-6

Svetina, D., & Rutkowski, L. (2017). Multidimensional measurement invariance in an international context: Fit measure performance with many groups. *Journal of Cross-Cultural Psychology, 48*(7), 991–1008. https://doi.org/10.1177/0022022117717028

Tan, J. J. X., Kraus, M. W., Carpenter, N. C., & Adler, N. E. (2020). The association between objective and subjective socioeconomic status and subjective well-being: A meta-analytic review. *Psychological Bulletin, 146*(11), 970–1020. https://doi.org/10.1037/bul0000258

van Ewijk, R., & Sleegers, P. (2010). The effect of peer socioeconomic status on student achievement: A meta-analysis. *Educational Research Review, 5*(2), 134–150. https://doi.org/10.1016/j.edurev. 2010.02.001

Yang, C., Luo, L., Vadillo, M. A., Yu, R., & Shanks, D. R. (2021). Testing (quizzing) boosts classroom learning: A systematic and meta-analytic review. *Psychological Bulletin, 147*(4), 399–435. https://doi.org/10.1037/bul0000309

Yang Hansen, K., & Gustafsson, J.-E. (2016). Causes of educational segregation in Sweden—School choice or residential segregation. *Educational Research and Evaluation, 22*(1–2), 23–44. https:// doi.org/10.1080/13803611.2016.1178589

Yang Hansen, K., & Gustafsson, J.-E. (2019). Identifying the key source of deteriorating educational equity in Sweden between 1998 and 2014. *International Journal of Educational Research.* https://doi.org/10.1016/j.ijer.2018.09.012

Yang Hansen, K., & Strietholt, R. (2018). Does schooling actually perpetuate educational inequality in mathematics performance? A validity question on the measures of opportunity to learn in PISA. *ZDM, 50*(4), 643–658. https://doi.org/10.1007/s11858-018-0935-3

Yang Hansen, K., Radišić, J., & Ding, Y. et al. (2022). Contextual effects on students' achievement and academic self-concept in the Nordic and Chinese educational systems. *Large-scale Assessments in Education, 10*(16). https://doi.org/10.1186/s40536-022-00133-9

Kajsa Yang Hansen's research concerns educational quality and equity from a comparative perspective. She tackles these issues by investigating students' social, motivational, and cognitive factors in the contexts of their schools, changing societies and education systems. Dr. Yang Hansen also has an interest in analytical techniques for large-scale survey data, e.g., multi-level analysis, Structural Equation Modelling (SEM) and second-generation SEM.

Victoria Rolfe (born 1987) is a senior lecturer at the Department of Education and Special Education, University of Gothenburg. Her research focus is on educational equity, educational measurement, and the opportunity to learn.

Nani Teig is an associate professor at the University of Oslo, Norway. Her research focuses on inquiry-based teaching, scientific reasoning, teaching quality, and academic resilience. She integrates multilevel analyses using data from videos, surveys, assessments, and computer log files. Dr. Teig has received several awards and fellowships, including the Global Education Award, Bruce H. Choppin Dissertation Award, Young CAS Fellow, and UNESCO GEM Fellow.

Chapter 10
Discussions of Findings on Teacher Practice Across Countries, Time, and Chapters

Trude Nilsen⬤ and Nani Teig⬤

10.1 Introduction

This chapter discusses the empirical chapters of the book. The overall structure of the book, as outlined in Chap. 1 and again shown in Fig. 10.1, organizes the empirical chapters according to the following three approaches:

1. *Examining the means over time and relations to student achievement* in 2019 for content coverage (Chap. 4) and teaching quality and assessment practices (Chap. 5);
2. *Explaining changes in achievement over time* for content coverage (Chap. 6) and teaching quality and assessment practices (Chap. 7); and
3. *Investigating the equality between different groups of students* with regard to content coverage (Chap. 8) and teaching quality and assessment practices (Chap. 9).

These approaches align with the main objectives of the book, which are to investigate:

1. how teacher practices have changed over time and their relations with student achievement in IEA's Trends in International Mathematics and Science Study (TIMSS) 2019,

T. Nilsen (✉)
Department of Teacher Education and School Research, Faculty of Educational Sciences, CREATE—Centre for Research on Equality in Education (project number 331640), University of Oslo, Postboks 1099, Blindern, 0317 Oslo, Norway
e-mail: trude.nilsen@ils.uio.no

N. Teig
Department of Teacher Education and School Research, University of Oslo, Postboks 1099, Blindern, 0317 Oslo, Norway
e-mail: nani.teig@ils.uio.no

© The Author(s) 2024
N. Teig et al. (eds.), *Effective and Equitable Teacher Practice in Mathematics and Science Education*, IEA Research for Education 14,
https://doi.org/10.1007/978-3-031-49580-9_10

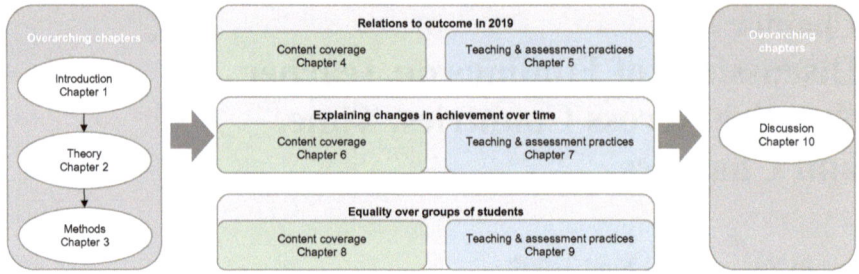

Fig. 10.1 Structure of the book according to the three empirical approaches

2. how changes in teacher practices are related to changes in achievement, and
3. how teacher practices are related to educational equality.

Alternatively, the book can be viewed from a more substantive or conceptual perspective, as illustrated in Fig. 10.2. This perspective categorizes the empirical chapters into two areas: (1) *content coverage* and (2) *teaching quality and assessment practices*. These two areas are examined from *three different perspectives*. For instance, in the area of content coverage, the authors examined:

1. how content coverage is related to student achievement in 2019 and how the mean of content coverage has changed over time (Chap. 4);
2. how changes in content coverage are related to the changes in achievement from 2011 to 2019 (Chap. 6), and
3. how content coverage is related to equality (Chap. 8).

In the exact same manner, these three approaches are undertaken to investigate teaching quality and assessment practices (Chaps. 5, 7, and 9).

Choosing whether to summarize and discuss the findings according to the empirical approaches (Fig. 10.1) or substantive perspective (Fig. 10.2) makes a significant difference. Discussing findings based on the three empirical approaches in Fig. 10.1 allows for an examination of the affordances and limitations of each specific approach. For instance, the approach of examining the relation between

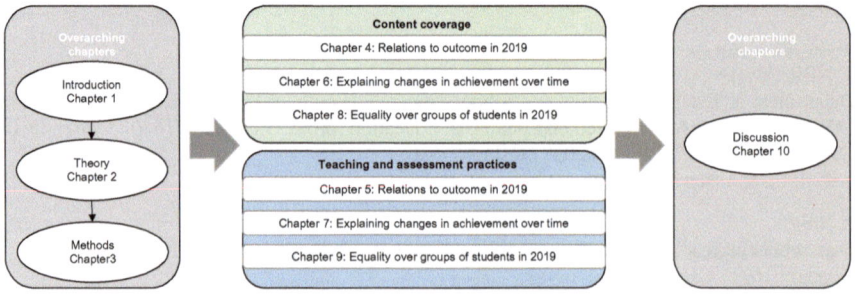

Fig. 10.2 Structure of the book from a substantive perspective

changes in teaching quality and changes in student achievement over time allows for more robust inferences. However, the constructs have changed over time, leading to reduced validity of the findings. Such discussions are methodologically focused and offer limited insights into the broader field of teacher practice.

On the other hand, summarizing, comparing, and discussing findings from a substantial perspective (Fig. 10.2) enables a more comprehensive exploration of teacher practice from many angles. This approach is valuable for policymakers, practitioners, and researchers. However, comparing findings from chapters that investigate the same concept with different approaches presents challenges. Comparability of findings may be jeopardized due to different operationalizations of the same concepts (e.g., the constructs may have changed over time) and the use of different analytical methods.

To account for the different considerations described above, this chapter starts with brief summaries of the findings from the empirical chapters. It then adopts the substantive perspective (Fig. 10.2) to compare and discuss findings related to content coverage and then teaching quality and assessment practices in light of previous research. Subsequently, the chapter discusses the reliability and validity of findings according to the three empirical approaches, as illustrated in Fig. 10.1. Lastly, the chapter outlines the limitations, contributions, and implications of findings from this book.

10.2 Short Summaries of the Findings of the Chapters

This section provides short summaries of the empirical chapters, following the structure of the book (Fig. 10.1) and the main aims of the book described in the previous section. For more details, please refer to the content of each chapter.

Chapter 4 investigates the alignment between the TIMSS test and the curricula of Nordic countries. It further investigates relations between content coverage and achievement, as well as to what extent the means of content coverage have changed over time. The findings show that there is indeed an alignment between the TIMSS test and the Nordic countries' curricula. However, few significant relations exist between the content coverage reported by teachers and student achievement. For instance, in 2019, significant positive associations were found between content coverage in geometry and achievement in geometry in Norway, as well as between content coverage in physical science and achievement in physical science in Sweden. Conversely, a negative relationship was found between content coverage in earth science and achievement in both earth science and physical science.

Chapter 5 examines the relationship between teaching quality and assessment practice with achievement using TIMSS data from 2019. It also investigates changes in the means of teaching quality and assessment practice from 2011 to 2019. The findings of the study suggest that classroom management is the strongest predictor of student achievement at the classroom level. Positive relationships were observed between teacher support and instructional clarity with student achievement in mathematics at the classroom level, while in science, these factors were found to be better predictors at the student level. There were no significant findings for cognitive activation, except for inquiry (measured as the frequency of conducting experiments) which had a curve-linear relationship with achievement. Homework frequency was positively related to mathematics achievement in Denmark and Sweden, while the time spent completing homework had a negative relationship with mathematics achievement in Finland. No significant findings were observed for teachers' assessment strategies. The study also revealed a decrease in teacher support and instructional clarity for all countries, except in Finland, with mixed results for cognitive activation. Homework assignments decreased in all countries, indicating that teachers allocated less homework in 2019 as compared to 2015 and 2011.

Chapter 6 explores the relationship between changes in content coverage and changes in achievement from 2011 to 2019. There were no significant findings for science. For mathematics, there were significant relations between changes in content coverage and changes in achievement for all countries, except for Sweden. Content coverage in topics for the number domain increased over time in all countries and was related to changes in achievement in Norway, Denmark, and Sweden. Content coverage in topics for geometry increased from 2011 to 2019 in Sweden and Finland but decreased in Norway. Changes in content coverage in topics for geometry were only significantly related to changes in achievement in Norway and Finland. Content coverage in topics for data did not change significantly over time in Denmark but decreased in Sweden and Finland and increased in Norway. However, these changes were only significantly related to changes in achievement in Finland and Norway. Due to the large amount of missing data in science, reliable and valid findings were not possible with regards to content coverage in science.

Chapter 7 investigates the relationship between changes in aspects of teaching quality and homework and changes in achievement from 2011 to 2019. Students' perceived instructional clarity decreased over time in all countries except for Finland. In mathematics, these changes were related to changes in mathematics achievement, while in science, this relation was only significant in Norway. Cognitive activation in science (measured by inquiry practice) decreased in Sweden and increased in Norway but did not change significantly for Sweden or Denmark. Changes in inquiry practice were related to increased achievements in science in Norway and Sweden. In mathematics, cognitive activation increased in Sweden and decreased in Norway, but these changes were not significantly related to changes in mathematics achievement. Limitations to teaching (e.g., students feel tired or hungry, students absent from class) exhibited a negative trend in all countries, indicating that teachers in 2019 reported on more challenges to teaching than in 2011. This negative change was related to changes in achievement over time in all Nordic countries. Homework decreased over time in most countries, but there were hardly any significant relations. Only in Sweden, there was a weak relation between changes in homework and changes in achievement.

Chapter 8 investigates the relationship between content coverage and inequality, specifically by examining whether content coverage mediates the relationship between SES and mathematics achievement. The findings reveal varied results among the Nordic countries. In Denmark, content coverage in the number and geometry topics positively related to SES, with a significant indirect effect in geometry. This indicates that content coverage exacerbates the effect of SES on achievement, thereby reducing equality. In Norway, there was a negative relation between data topics and SES, indicating possible compensatory approaches by schools and increased equality. In Finland, there was no relation between SES and content coverage, and thus no mediation effect occurred. There were no significant effects in Sweden. Furthermore, in Norway and to a certain degree in Sweden, there is a decrease in the amount of content coverage among high-SES students, indicating a compensatory strategy that comes at the expense of students with higher SES backgrounds.

Chapter 9 analyses the relationship between teacher practice and inequality. It further examines whether socioeconomic and ethnic inequalities in students' mathematics achievement differ significantly across classrooms. The findings reveal two main points. Regarding teacher practice, few significant findings were found, with only Norway exhibiting a relationship between teacher practices and equality. High SES classrooms in Norway reduced inequalities through teachers' emphasis on academic success and assessment practices. In Denmark, classrooms with a high SES composition were positively associated with teachers' emphasis on academic success, which in turn, was positively associated with a supportive climate. This was true also for Finland, except that teachers' emphasis on academic success not only impacted a supportive climate, but also cognitive activation. In Sweden, classrooms with a high SES composition and few minority students, were positively related to teachers' emphasis on academic success, which in turn, was positively related to a supportive climate and cognitive activation. Furthermore, cognitive activation was positively associated with high content coverage. With regards to socioeconomic and ethnic inequalities, the findings show that classroom socioeconomic and ethnic contexts significantly impact students' mathematics achievement. Attending classrooms with a high proportion of students with high socioeconomic status was found to have a compensatory effect in reducing the effect of family socioeconomic status on students' achievement in Norway and Finland. Similarly, attending classrooms with a high proportion of native students was associated with a smaller effect of student ethnicity affecting their achievement in Denmark. No such effects were identified in Sweden.

10.3   Discussion of Findings on Content Coverage

In this section, we discuss the findings on teaching quality and assessment practices by following the substantive perspective illustrated in Fig. 10.2. We focus on three chapters addressing these aspects of teacher practices: Chaps. 4, 6, and 8. We will synthesize the findings from these chapters, compare them across chapters, and relate them to previous research. This section emphasizes the discussion on how content coverage:

- has changed over time,
- is related to student achievement,
- is related to the changes in achievement over time, and
- is related to educational equality.

Changes in means over time. Chapters 4 and 6 investigate how the means of content coverage in different topics of mathematics and science have changed over time. Both chapters find that there is a larger percentage of students who had teachers who covered the topic number in 2019, compared to 2011. This pattern is consistent across the chapters and across countries. In geometry topics, there was a positive trend for Sweden and Finland, a negative trend for Norway, and no significant findings for Denmark. For data topics, there was a negative trend for Sweden and Finland, a positive trend for Norway, and no significant findings for Denmark. In other words, the only clear pattern identified across countries was that of the positive trend for the topic number. For science, there are no consistent patterns of changes over time in the content coverage of life science, physical science, or earth science across the Nordic countries.

Relations to achievement. Chapters 4, 6, and 8 include investigations of the relations between content coverage and achievement. In Chap. 4, there were few significant findings. In Chap. 6, the data from the three cycles were merged, revealing positive relations between content coverage in all mathematics topics (number, geometry, and data) and student achievement. The relations were small, and significant across all countries except for Sweden. Using data from TIMSS 2019 and controlling for SES, Chap. 8 found positive relations between content coverage in number topics and student achievement in mathematics in all countries except for Sweden. Positive relations were also found for content coverage in geometry and data, but only for Denmark.

Taken together, the findings indicate that: (1) the relations between content coverage and achievement vary across countries, with Sweden being an outlier, and (2) the construct of content coverage has a low statistical power. Only when data from all three cycles are merged, are the relations between content coverage and achievement significant (except for Sweden). This is likely due to the fact that the construct is based on teachers' responses, and the sample size of teachers is not representative and is also quite small. Another possible explanation is the restriction of time range since within a country in a single cycle, it is rarely expected for teachers to have huge variations in content coverage. The low statistical power could also be due to the low reliability and/or validity of the construct (see Sect. 10.5). Most probably, the reason for the many insignificant findings when using one cycle only is a mix of both. The relation between content coverage and achievement should, according to previous research, be positive and significant (Scheerens, 2016). Indeed, the assumption behind effective learning, is that students' opportunity to learn the content should have a positive effect on their learning outcome (Scheerens, 2016; Schmidt et al., 2015).

Explaining changes in achievement over time. Chapter 6 investigates the relations between changes in content coverage and changes in achievement. The findings indicate that changes in content coverage in number, geometry, and data, were related to the changes in mathematics achievement. These indirect effects were small, and significant in all countries except for Sweden. Previous research indicates that the changes in students' opportunity to learn the content will affect their learning outcomes (Scheerens, 2016). Hence, the findings from Sweden are not in line with previous research, and more research is needed to identify the reasons behind this discrepancy.

Relations to educational equality. Chapter 8 investigates how content coverage is related to equality. Of all the Nordic countries, Denmark stands out. In Denmark, high-SES classrooms were to a higher degree exposed to the content in all topics of mathematics. This implies that students in these classrooms were provided with a better opportunity to learn than those in low-SES classrooms. In geometry, there was a significant mediation effect, meaning that the content coverage added to the effect of students' socioeconomic background. Content coverage in Denmark, hence, increased the gap between low- and high- SES students. In contrast, the opposite was found in Norway. Norwegian students from lower socioeconomic backgrounds were more often exposed to data topics than students from higher SES backgrounds. A similar tendency was found in Sweden for the number topics. These results may point to compensatory approaches by schools and teachers to provide additional opportunities for disadvantaged students.

These findings are partly in line with previous research, which found that content coverage might add to the existing inequalities caused by students' home background in several countries (Schmidt et al., 2015). Schools may either exacerbate or reduce existing inequalities through students' opportunities to learn the content of the subject domain.

10.4 Discussion of Findings on Teaching Quality and Assessment Practices

In a similar fashion, as in the previous discussion on content coverage, this section discusses how teaching quality and assessment practices:

- have changed over time,
- are related to achievement,
- are related to changes in achievement over time, and
- are related to equality.

Changes in means over time. Chapters 5 and 7 investigate whether teaching quality and assessment practices have changed over time, albeit using different methods of analysis. Both chapters find that teacher support and instructional clarity reported by students decreased from 2011 to 2019. This suggests that students perceive their teachers provide less support and less clear and understandable instruction in 2019 than in 2015 and 2011. These results were consistent across countries (except for Finland) and across subject domains (mathematics and science). There were no clear patterns for cognitive activation. Homework assignments decreased over time. The measures for other aspects of teacher practices had changed over time, making it impossible to compare the means of these across the three cycles.

It is challenging to determine why teacher support and clarity decreased over time and whether it was in fact the teaching quality that changed or whether it was students' *perceptions* of their teachers that changed. As pointed out in Chap. 5, there are indications that the student composition may have changed, and additionally, teachers reported that there were more challenges or factors limiting their teaching in 2019 as opposed to 2011. Teachers today do indeed face more challenges, as teaching in heterogeneous classrooms where many minority students struggle with language difficulties, is more demanding than teaching homogenous classrooms.

The findings that teachers assigned less homework in 2019 as opposed to 2015 and 2011 raise intriguing questions and open up new areas of investigation. Uncovering the factors driving this trend is not straightforward and calls for further in-depth research.

Relations to achievement. Chapters 5 and 7 explore the relations between teaching quality and assessment practices with student achievements. A pattern of positive relations between teaching quality and achievement is evident across these chapters. However, among the different aspects of teaching quality, classroom management stands out with a strong and consistent relation to achievement. This finding is in line with previous research (e.g., Charalambous & Praetorius, 2020; Senden et al., 2023). Teacher support was found to be positively associated with higher achievement in mathematics, while there were few positive findings in science. There were mostly insignificant findings for cognitive activation, except for inquiry, which is an aspect of cognitive activation in science. The positive relation between inquiry and achievement has been confirmed in previous research (e.g. Teig et al., 2018, 2021, 2022). With the exception of inquiry, cognitive activation seems to be more important in secondary school than in primary school (Nilsen et al., 2018). However, longitudinal data is needed to confirm this.

There were very few significant findings between assessment practices (including homework) and achievement in any of the chapters. Indeed, findings from previous research on this topic are mixed, especially for homework (Fernández-Alonso & Muñiz, 2021). Further research with more nuanced data going deeper into assessment practices and homework is required.

Explaining changes in achievement over time. Chapter 7 investigates whether changes in teaching quality and assessment practices (homework) are related to changes in achievement. There are weak but significant relations between changes in teacher support and changes in student achievement, with the pattern being clearer and stronger in mathematics. Additionally, there was further a pattern of weak but significant relations between changes in inquiry practices and changes in student achievement in science. However, a much stronger and persistent pattern was found for limitations to teaching (e.g., students feel tired or hungry). The findings indicated that limitations to teaching seemed to either hinder further increased achievement for the countries where achievements increased from 2011 to 2019 or explain decreased achievements for the countries with negative trends in achievement during the same period. For Norway, this finding is aligned with previous research on students in grade nine (e.g. Kaarstein & Nilsen, 2021; Nilsen et al., 2022; Vik et al., 2022). However, further research is needed in the other Nordic countries to confirm this finding, preferably with longitudinal data.

When examining teachers' assessment practices, there were no significant patterns or consistent relationships. This means that the chapter did not find a clear link or correlation between the changes in achievement and the changes in assessment practices that focus on homework frequency, homework time, and in-class homework discussion.

Relations to educational equality. Chapter 9 investigates the relations between teacher practice and educational inequalities. The findings suggest that in all Nordic countries, except Norway, teacher practices neither reduce nor increase the relation between SES and achievement. However, in Norway, being a minority student had less impact on achievement in high SES classrooms than low SES classroom. According to previous research, teaching quality has the potential to reduce or increase inequalities in learning outcomes caused by SES (Cardichon et al., 2020; Nilsen et al., 2020; Rjosk et al., 2014). Hence, the insignificant findings for the other Nordic countries are to some degree unexpected. The few significant findings could be related to the complexity of the statistical model used. This model includes a large number of variables, and it includes classroom SES, which is a powerful predictor. In this chapter, classroom SES and minority background are the main moderators of the relations between individual SES (and minority) on achievement, mediated via teacher practices, school emphasis on success (SEAS), and opportunities to learn (OTL). In other words, the authors examine whether classroom SES and minority may reduce achievement gaps between individual students (via teacher practices, SEAS, and OTL). If the research question rather focused on teacher practices as the moderator of the relations between individual SES (and minority) on achievement (controlled for SEAS and OTL), this could have provided more information on whether teacher practices may reduce inequalities.

Chapter 9 also includes findings on relations between classroom composition and teacher practice. Apart from Norway, high SES classrooms were positively associated with high values of teachers' emphasis on academic success which in turn was positively associated with higher teaching quality in the other Nordic countries. This could indicate unequal distribution of high-quality teachers, in that high-quality teachers tend to teach high SES schools. This is in line with previous research (Cardichon et al., 2020; Nilsen & Bergem, 2020; Qin & Bowen, 2019), and an indication of inequality in access to good teachers.

10.5 Reliability and Validity

In our discussions about the reliability and validity of findings, we follow the structure of the book, as illustrated in Fig. 10.1. This means that the discussions are structured according to the three empirical approaches used in the book:

- **Approach 1**: analyses of relations between teacher practices and student achievement in 2019, including estimations of means of teachers practices across the three cycles of TIMSS.
- **Approach 2**: analyses of relations between changes of teachers practices with changes in achievement from 2011 to 2019.
- **Approach 3**: analyses of the interplay between teacher practices and educational inequality.

10.5.1 Approach 1: Relations to Achievement in 2019

Chapters 4 and 5 both utilized two-level structural equation modelling at the students and classroom levels to investigate relations between teacher practice and student achievement in 2019. This analytical method allows for simultaneous estimations at both levels and provides model fits (Hox & Roberts, 2011). Structural equation modelling, using the software Mplus, further provides the reliability and validity of the constructs through confirmatory factor analyses, yielding robust findings (Morin et al., 2014).

In 2019, the validity of teaching quality was high (Klieme & Nilsen, 2022). This can be attributed to several factors. The measures for teaching quality in TIMSS are heavily based on theory and previous research (Baumert et al., 2010; Klieme et al., 2009; Mullis & Martin, 2017). Since 1995, TIMSS has already included aspects of teaching quality (Klieme & Nilsen, 2022). Building on this foundation, a broader and more comprehensive measure of teaching quality was introduced in TIMSS 2015 and has been piloted, tested, and improved several times.

The reliability and validity of content coverage, on the other hand, are more problematic. This is first and foremost related to the response scale. Teachers are asked when the students in the class were taught various topics within the domains of geometry, number, and data in the subject of mathematics. Similarly, they are asked about all the topics within the domain's life science, physical science, and earth science in the subject of science. The response scale consists of three categories: *mostly taught before this year*, *mostly taught this year*, and *not yet taught or just introduced*.

The problem with the response scale lies primarily in the response option "not yet taught or just introduced", as this could be interpreted to include two different responses (not yet taught/just introduced). This ambiguity may be the reason why

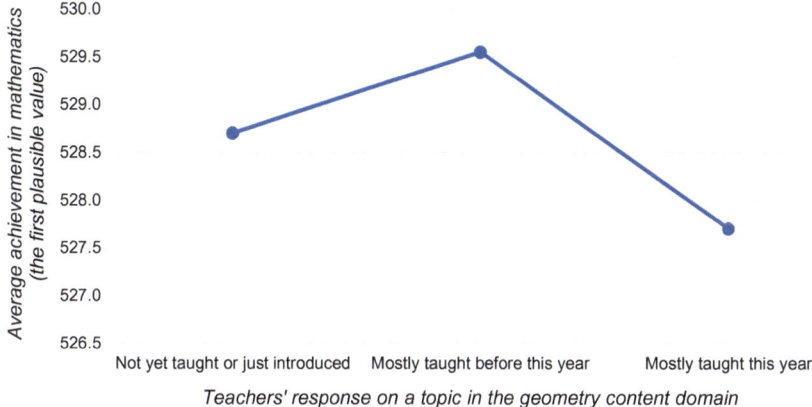

Fig. 10.3 The graph association between teachers' response on a topic in the domain geometry and the first plausible value of mathematics achievement

higher values of the response scale are not necessarily associated with higher achievement, as illustrated in Fig. 10.3. In this figure, "mostly taught before this year" corresponds to higher achievement than "not yet taught or just introduced". This makes sense, however, one would expect "mostly taught this year" to correspond to the highest achievement compared to other responses. In TIMSS 2023, this issue has been addressed.

Another factor contributing to the low validity of content coverage is the large number of missing data, especially in science. The exact reasons for the high number of missing values are not clear, but it is possible that survey fatigue played a role, as the questions about science topics come after those about mathematics topics. Additionally, there are numerous items for the teachers to complete, which could also contribute to missing data due to fatigue. These issues may have contributed to the small number of significant findings in the chapters examining content coverage, along with the small and not representative sample of teachers.

10.5.2 Approach 2: Explaining Changes in Achievement

Approach 2 is the analyses of the relations between changes in teachers' practice and changes in achievements and are employed in Chaps. 6 and 7. The method of analyses used in these two chapters is robust and increases the plausibility of causal inferences. In this respect, the inferences should be reliable and valid. However, several of the constructs have changed over time, which may reduce the reliability and validity of some results. For instance, the aspect of teaching quality referred to as cognitive activation has changed from 2011 to 2019. The following six items were included in both cycles 2019 and 2015, but not in 2011: *ask students to explain their answers*;

ask students to complete challenging exercises that require them to go beyond the instruction; encourage classroom discussions among students; link new content to students' prior knowledge; ask students to decide their own problem-solving procedures; and encourage students to express their ideas in class. For cognitive activation in mathematics, only two items remained unchanged from 2011 to 2019, reducing the validity of the construct.

Regarding content coverage, there were few changes over time (for more on this, see Chap. 6). However, the large amount of missing data in science (except for Finland) across several cycles prevented reliable and valid findings concerning content coverage in science.

10.5.3   Approach 3: Equality

Approach 3 involves analyses designed to disentangle the interplay between teacher practice and educational equality. Chapters 8 and 9 investigate this for content coverage and teacher practices, respectively. However, the two chapters use different approaches.

Chapter 8 uses an additive approach, where content coverage mediates the relation between SES and achievement. This approach is referred to as *additive* because it investigates whether a mediator (in this case, content coverage) may *add* to the effect of the predictor (here, SES). This approach is in accordance with previous theories on opportunities to learn and equality (e.g. Schmidt et al., 2015).

On the other hand, Chap. 9 uses a differential approach and investigates whether teacher practice *moderates* the relation between SES and achievement (or the effect of ethnicity on achievement). This approach is used to investigate whether a moderator (in this case, teacher practice) may reduce or increase (i.e., moderate) the strength of the effect of SES on achievement. This approach is in accordance with previous research (e.g. Gustafsson et al., 2018; Nilsen et al., 2020), as one may expect, different aspects of teacher practice have varying effects on different groups of students (hence the term *differential effect*). However, in Chap. 9, this moderation is just part of a larger model that investigates whether attending a school or classroom with high SES reduces or increases the importance of students' individual home background via teacher practice.

While both chapters use robust and advanced methodology, they report few significant findings, which is partly contradictory to previous research. The limited significant findings in Chap. 8 may be due to the low validity of the content coverage construct, while the few significant findings in Chap. 9 might be attributed to the complex model incorporating a large number of variables and relationships.

10.6 Contributions and Concluding Remarks

The main aim of this book was to investigate teacher practices in grade four from multiple perspectives: how they are related to achievement, how they have changed over time and whether these changes are related to the changes in achievement, and how they are related to equality. The summaries of the findings from all the empirical chapters and the discussions of these findings across chapter and in light of previous research has revealed:

(1) The findings are mostly in line with previous research.
(2) Teaching quality is crucial to student achievement. In primary school (grades four and five), the results indicate that classroom management in mathematics, inquiry practices in science, as well as teacher support and instructional clarity in both subjects are associated with higher learning outcomes.
(3) Students' opportunity to learn the content in mathematics (content coverage) is essential to student achievement; however, results vary across countries and content domains (number, geometry, and data)
(4) Teachers in Nordic countries have reduced the amount of homework they assign to students over time, and findings on homework are mixed. More research is needed with data providing more in-depth information, including the type of homework assigned to the students (e.g., focus on skill development or content reinforcement), and how these different types may affect student outcomes.
(5) Students in high SES classrooms or schools are provided with better opportunities to learn the content (especially in Denmark) and experience higher teaching quality. However, in Norway, teaching quality compensates the gap between disadvantaged and advantaged students.
(6) Teachers are facing growing challenges in their teaching due to the heterogeneity of their classrooms and students who may not have had sufficient sleep or nutrition. These challenges to teaching may explain the decreased achievement observed or hinder further increases in achievement since 2011.

This book contributes to educational policy and points to the need for more equal opportunities for students to learn. It is problematic that high SES schools tend to get the best teachers, as this exacerbates existing inequalities caused by students' home backgrounds and minority status. This contradicts the Nordic model of education (see Chap. 1). In our times of growing inequalities caused by the COVID-19 pandemic, war, climate and energy crises, the inequalities in society, spread and influence schools as well. It is therefore important to decrease inequalities in society and equip teachers to teach in heterogeneous classrooms. A substantial body of research has shown that equality promotes quality in school and prosperous societies (Hanushek & Woessmann, 2015; Wilkinson & Pickett, 2011).

This book also contributes to teacher education and the research in the fields of teaching quality and teacher effectiveness. Most research on teacher practices comes from the USA or Germany, and research on teaching quality in mathematics and science more often involves students in secondary school rather than students in grade four or grade five. Our research adds to this previous research by including the Nordic context and younger students.

In terms of methodology, our book has the advantage of using the same concepts and operationalizations, methodological approaches, and the same samples and data (see Chap. 3). This allows for comparisons of findings across chapters and facilitates a deeper understanding of the relationships between variables. The methods of analysis employed in the various chapters have been carefully chosen to ensure robust, reliable, and valid inferences, which are essential for the credibility and impact of the research findings. Moreover, the method used to examine the relations between changes in teacher practice and changes in achievement could prove beneficial for other countries as well, as it provides a replicable and adaptable framework for analyzing similar phenomena in different contexts.

A mantra followed since 1995 in TIMSS is: *if you want to measure change, don't change the measure* (Martin et al., 2020). It is indeed important to follow this, as evidenced by the challenges faced by the chapters that investigated changes in teacher practice in relation to changes in achievement (Chaps. 6 and 7). For the most part, TIMSS does indeed follow this rule. However, there needs to be a balance between maintaining trend measures and the need to improve and align with changes by the societies and schools of the participating countries. In the case of teaching quality, the change has improved the quality of the measures and enabled research of this important concept (Klieme & Nilsen, 2022).

The classroom design of TIMSS is particularly valuable for examining the role of teacher practice in relation to students' learning outcomes and educational equality. By sampling intact classes from schools, and collecting representative data from students, their teachers, and parents through questionnaires, TIMSS offers a comprehensive and unique perspective on the educational landscape. This approach sets TIMSS apart from other international large-scale studies, such as the Teaching and Learning International Survey (TALIS) and the Programme for International Student Assessment (PISA). TALIS does not measure student learning outcomes, while PISA does not sample whole classes, include teacher questionnaires for the teachers of the sampled students, or involve primary school students.

The wealth of data provided by TIMSS has enabled us to draw meaningful comparisons about effective and equitable teacher practice across the Nordic countries. By sharing insights and best practices, these countries can work towards a more unified and effective Nordic education system, while reducing disparities and inequalities that have emerged over time. Hopefully, our book contributes to this ongoing dialogue and takes us one step closer to achieving the Nordic model's aim of greater educational equity and excellence.

References

Baumert, J., Kunter, M., Blum, W., Brunner, M., Voss, T., Jordan, A., Klusmann, U., Krauss, S., Neubrand, M., & Tsai, Y.-M. (2010). Teachers' mathematical knowledge, cognitive activation in the classroom, and student progress. *American Educational Research Journal, 47*(1), 133–180. https://doi.org/10.3102/0002831209345157

Cardichon, J., Darling-Hammond, L., Yang, M., Scott, C., Shields, P. M., & Burns, D. (2020). *Inequitable opportunity to learn: student access to certified and experienced teachers.* Learning Policy Institute. https://learningpolicyinstitute.org/product/crdc-teacher-access

Charalambous, C. Y., & Praetorius, A.-K. (2020). Creating a forum for researching teaching and its quality more synergistically. *Studies in Educational Evaluation, 67*, 100894. https://doi.org/10.1016/j.stueduc.2020.100894

Fernández-Alonso, R., & Muñiz, J. (2021). Homework: facts and fiction. In: Nilsen, T., Stancel-Piątak, A., & Gustafsson, J. E. (Eds.), *International handbook of comparative large-scale studies in education.* Springer. https://doi.org/10.1007/978-3-030-38298-8_40-1

Gustafsson, J.-E., Nilsen, T., & Hansen, K. Y. (2018). School characteristics moderating the relation between student socio-economic status and mathematics achievement in grade 8. Evidence from 50 countries in TIMSS 2011. *Studies in Educational Evaluation, 57*, 16–30. https://doi.org/10.1016/j.stueduc.2016.09.004

Hanushek, E. A., & Woessmann, L. (2015). *The knowledge capital of nations: Education and the economics of growth.* MIT Press.

Hox, J., & Roberts, J. K. (Eds.) (2011). *Handbook of advanced multilevel analysis.* Psychology Press.

Kaarstein, H., & Nilsen, T. (2021). Lærerkompetanse, undervisningskvalitet og naturfagprestasjoner fra TIMSS 2015 til TIMSS 2019 [Teacher competence, teaching quality, and science performance from TIMSS 2015 to TIMSS 2019]. In *Med blikket mot naturfag: Nye analyser av TIMSS 2019-data og trender 2015–2019 [Focused on science: new analyses of TIMSS 2019 data and trends from 2015–2019]* (pp. 183–206). Universitetsforlaget. https://doi.org/10.18261/978821 5045108-2021-0

Klieme, E., & Nilsen, T. (2022). Teaching quality and student outcomes in TIMSS and PISA. In T. Nilsen, A. Stancel-Piątak, & J. E. Gustafsson (Eds.), *International handbook of comparative large-scale studies in education* (Vol. 2). Springer. https://doi.org/10.1007/978-3-030-38298-8_37-1

Klieme, E., Pauli, C., & Reusser, K. (2009). The pythagoras study: Investigating effects of teaching and learning in Swiss and German mathematics classrooms. In *The power of video studies in investigating teaching and learning in the classroom* (pp. 137–160). Waxmann.

Martin, M. O., von Davier, M., & Mullis, I. V. S. (2020). *Methods and procedures: TIMSS 2019 technical report.* TIMSS & PIRLS International Study Center, Boston College. https://timssa ndpirls.bc.edu/timss2019/methods

Morin, A. J. S., Marsh, H. W., Nagengast, B., & Scalas, L. F. (2014). Doubly latent multilevel analyses of classroom climate: An illustration. *The Journal of Experimental Education, 82*(2), 143–167. https://doi.org/10.1080/00220973.2013.769412

Mullis, I. V. S., & Martin, M. O. (2017). *TIMSS 2019 assessment frameworks.* TIMSS & PIRLS International Study Center, Boston College. http://timssandpirls.bc.edu/timss2019/frameworks/

Nilsen, T., & Bergem, O. K. (2020). Teacher competence and equity in the Nordic countries. mediation and moderation of the relation between SES and achievement. *Acta Didactica Norden, 14*(1). https://doi.org/10.5617/adno.7946

Nilsen, T., Kaarstein, H., & Lehre, A.-C. (2022). Trend analyses of TIMSS 2015 and 2019: School factors related to declining performance in mathematics. *Large-Scale Assessments in Education, 10*(1), 15. https://doi.org/10.1186/s40536-022-00134-8

Nilsen, T., Scherer, R., & Blömeke, S. (2018). 3. The relation of science teachers' quality and instruction to student motivation and achievement in the 4th and 8th grade: A Nordic persepctive. *Northern Lights on TIMSS and PISA 2018*, 61. Nordic Council of Ministers.

Nilsen, T., Scherer, R., Gustafsson, J.-E., Teig, N., & Kaarstein, H. (2020). Teachers' role in enhancing equity: A multilevel structural equation modeling with mediated moderation. In T. S. P. Frønes, Andreas; Radišić, Jelena & Buchholtz, Nils (Eds.), *Equity, equality and diversity in the Nordic model of education*. Springer. https://doi.org/10.1007/978-3-030-61648-9_7

Qin, L., & Bowen, D. H. (2019). The distributions of teacher qualification: A cross-national study. *International Journal of Educational Development, 70*, 102084. https://doi.org/10.1016/j.ije dudev.2019.102084

Rjosk, C., Richter, D., Hochweber, J., Lüdtke, O., Klieme, E., & Stanat, P. (2014). Socioeconomic and language minority classroom composition and individual reading achievement: The mediating role of instructional quality. *Learning and Instruction, 32*, 63–72. https://doi.org/10.1016/j.learninstruc.2014.01.007

Scheerens, J. (2016). Opportunity to learn, curriculum alignment and test preparation: A research review. *Springer*. https://doi.org/10.1007/978-3-319-43110-9

Schmidt, W. H., Burroughs, N. A., Zoido, P., & Houang, R. T. (2015). The role of schooling in perpetuating educational inequality: An international perspective. *Educational Researcher, 44*(7), 371–386. https://doi.org/10.3102/0013189X15603982

Senden, B., Nilsen, T., & Teig, N. (2023). The validity of student ratings of teaching quality: Factorial structure, comparability, and the relation to achievement. *Studies in Educational Evaluation, 78*, 101274. https://doi.org/10.1016/j.stueduc.2023.101274

Teig, N., Bergem, O. K., Nilsen, T., & Senden, B. (2021). Gir utforskende arbeidsmåter i naturfag bedre læringsutbytte? [Does inquiry-based teaching practice in science provide better learning outcomes?]. In T. Nilsen & H. Kaarstein (Eds.), *Med blikket mot naturfag: Nye analyser av TIMSS 2019-data og trender 2015–2019 [Focused on science: new analyses of TIMSS 2019 data and trends from 2015–2019]* (pp. 46–72). Universitetsforlaget. https://doi.org/10.18261/9788215045108-2021-03

Teig, N., Scherer, R., & Nilsen, T. (2018). More isn't always better: The curvilinear relationship between inquiry-based teaching and student achievement in science. *Learning and Instruction, 56*, 20–29. https://doi.org/10.1016/j.learninstruc.2018.02.006

Teig, N., Scherer, R., & Olsen, R. V. (2022). A systematic review of studies investigating science teaching and learning: Over two decades of TIMSS and PISA. *International Journal of Science Education, 44*(12), 2035–2058. https://doi.org/10.1080/09500693.2022.2109075

Vik, F. N., Nilsen, T., & Øverby, N. C. (2022). Associations between sleep deficit and academic achievement—Triangulation across time and subject domains among students and teachers in TIMSS in Norway. *BMC Public Health, 22*(1), 1790. https://doi.org/10.1186/s12889-022-141 61-1

Wilkinson, R., & Pickett, K. (2011). *The spirit level: Why greater equality makes societies stronger*. Bloomsbury Publishing.

Trude Nilsen is a research professor at the University of Oslo. She is a leader of Strand 1 at CREATE - Centre for Research on Equality in Education, a leader of the research group LEA, and of funded research projects. She has been engaged as an international external expert for IEA's TIMSS and for OECD's TALIS studies. Her research focuses on teaching quality, educational equality, school climate, and applied methodology including causal inferences.

Nani Teig is an associate professor at the University of Oslo, Norway. Her research focuses on inquiry-based teaching, scientific reasoning, teaching quality, and academic resilience. She integrates multilevel analyses using data from videos, surveys, assessments, and computer log files. Dr. Teig has received several awards and fellowships, including the Global Education Award, Bruce H. Choppin Dissertation Award, Young CAS Fellow, and UNESCO GEM Fellow.